D0886894

Lesser sandhill cranes on a roosting site in the Platte near Grand Island. Photo by author

Paul A. Johnsgard

The Platte

Channels in Time

University of Nebraska Press: Lincoln & London

Manufactured in the
United States of America
The paper in this book
meets the guidelines for
permanence and durability
of the Committee on
Production Guidelines for
Book Longevity of the
Council on Library Resources.

Library of Congress
Cataloging in Publication Data

Johnsgard, Paul A.
The Platte: channels in time.

Includes bibliographical
references and index.
1. Stream ecology—
Nebraska—Platte River.
2. Platte River (Neb.)
I. Title.
QH105.N2J63 1984
574.5'26323'09782
83-10453
ISBN 0-8032-2560-1

In memory of the many

who have died along its banks;

In tribute to the few

who are striving to protect it;

And in uncertain hope

that those still to come will be

able to know it as I do.

Contents

Illustrations

Aerial view of Shoemaker Island. Photo courtesy Conservation and Survey Division, University of Nebraska

Preface

Love for the Platte River is something like deep appreciation for the works of Thomas Hart Benton; it does not come immediately. There are no towering mountains to lift the spirit, and there are no raging cataracts or waterfalls to overwhelm the ear and eye. Instead, there is a sense of familiarity, of immediate if superficial recognition, and of simple middle-class Americanism about the scene. Both the Platte and Benton's works provide a kind of popular respectability and lack of ostentation that offer a strong sense of association and security, but not of unrestrained excitement or grandeur.

This is not to say that the Platte River and the art of Benton are not exciting, dynamic, or complex, but all these features must be carefully searched for. They do not intrude on the consciousness like a Grand Canyon by Thomas Moran or a Matterhorn by Albert Bierstadt. The Platte's complexity and dynamism are thus all the more to be savored and individually enjoyed; it is perhaps too easy to be overwhelmed by mountains and canyons and to forget to search for their inner beauty and values beyond the obvious.

Coming to Nebraska and seeing the Platte River for the first time more than twenty years ago was like revisiting my childhood; the Red River of the North and the Platte River superficially have much in common. But the Platte is a river system rich in human and natural history to a degree that the Red River can hardly match, and my spirit is never raised by the Red River as it is by the Platte River each time I make a pilgrimage to its banks.

It was this slowly growing sense of awe for the Platte and its unique history that motivated me to commit my impressions to paper, and to hope that doing so might help other people to appreciate better the central role the Platte has played in shaping Nebraska and its heritage, both human and natural. This is by no means an exhaustive account of the riches of the Platte River and its valley, but a brief and personal portrait of the river as it has existed in the past and still exists today. My special hope in doing so is that the value of the river as a living, complex, and natural ecologic system can be more fully appreciated, and that whatever steps may be necessary might be taken to preserve the river's integrity. Otherwise, we risk the very real possibility that only a few decades from now, the river, like the vast bison herds that once trod its banks, will exist only as a remnant.

Although the primary emphasis of the book is on the present-day ecology of the river and its wildlife, especially in the vicinity of Grand Island, I have attempted to place the river in a much broader context of time and space, making brief mention of some of the smaller tributary valleys and adjoining river basins, such as the Niobrara to the north and the Republican to the south. I have also carried the story of the Platte forward from its earliest beginnings in prehistoric times to the present, and even slightly beyond, as least as far as one can safely predict the future with any degree of confidence. Predicting the future is by no means as comfortable as resurrecting the past, and I hope that any reader who chances upon this book in the early twenty-first century and is able to find the Platte River only as a shrunken stream will be tolerant, both toward the writer and especially toward his fellow humans for having allowed the river to be destroyed.

Adult male bison. Photo by author

There where the stream
of water separates,
There on the island's point,
where stands a tree clump
Where once great herds
of buffalo started from,
There I stood.
I stood surprised and rejoiced,
I stood surprised and rejoiced,
For there were the buffalo, like
threads covering the earth.
There they were, there they were.

Pawnee song (Densmore, 1929)

The Past

A red fox pup peering out of its den in late spring. Photo by author

The expanse of the world is wide,
My brother the fox spoke and said,
"Behold and see the wideness of the earth,
The white foxes know the earth is wide."

Pawnee song (Densmore, 1929)

The Land

For nearly a hundred million years the area that is now Nebraska lay beneath the shallow Mesozoic seas, slowly accumulating sedimentary deposits of the late Jurassic and Cretaceous periods and interring within these deposits the invertebrate and lower vertebrate life of that archaic era. At about the same time the earliest mammals were venturing out onto the land areas farther to the west, in what is now Wyoming, and encountering the last great reptiles of the time, the dinosaurs and their kin. By late Cretaceous times, western Nebraska was still covered by shallow seas. Along beaches and tidal-flat areas, sandstones and similar sedimentary rocks were laid down in the area now crossed by the North Platte River in the Ogallala vicinity, although they now lie deeply buried by deposits of more recent age.

At the end of the Cretaceous period, some 65 million years ago, vast changes occurred in the climate and land areas of the world. Whether these changes occurred fairly slowly or as a result of some vast cataclysm is still uncertain, but in any case it is known that of almost fifty families of reptiles known to be alive in Cretaceous times, half disappeared at the end of that period. This great reptilian die-off included all of the terrestrial dinosaurs, the winged pterosaurs, and the aquatic ichthyosaurs and plesiosaurs. Evidently no land animals larger than about fifty pounds survived, and among the reptiles only the turtles, lizards, snakes, and some aquatic alligatorlike forms survived to enter the Cenozoic era, the Age of Mammals.

At the beginning of the new age there may have been a resurgence of mountain building and volcanic activity that affected the climate of the world by releasing volcanic dust into the atmosphere, or perhaps even an asteroid or comet had a cataclysmic encounter with the earth. If that were true, a shockwave might have been generated that was great enough to kill instantly all of the larger land-dwelling animals and also destroy most of the forests of the earth. Additionally, great clouds of dust might have been sent into the atmosphere, shutting off light from the sun and perhaps changing not only the earth's climate, but also indirectly killing much of the plant life on which the land animals of the world depended. We may never know exactly what happened, but in any case the scene was quickly changed from a reptile-dominated terrestrial world to one ready to be taken over by the small and warm-blooded birds and mammals, which until that time had literally lived in the shadows of the larger dinosaurs and their kin.

With the early stages of the Cenozoic era, which began about 65 million years ago, the area now represented by Nebraska began to be molded by the accumulation of materials carried into the region from the mountains to the west. First, a deep wedge-shaped layer of sediment, now known as the Brule formation of the White River group, spread out from the mountains into western Nebraska. Next, additional layers of materials were laid on top of it, including the Arikaree deposits of Miocene times. On top of these were deposited the Ogallala sediments, which extended all the way from Nebraska's

CHART I: Time-scale of Cenozoic events in North America, with special reference to Nebraska and the Platte Valley.

Period	Epoch	Years before Present	Period or Stratum	Nebraska Examples
			Archaeologic Phase or Tradition	*Local Associated Culture*
		0–150	Modern	European
		150–250	Historic	Historic Pawnee
	Holocene	250–450	Late Ceramic	Dismal & Lower Loup
		2–4,000	Early Ceramic	Plains–Woodland
		4–9,000	Archaic	Hunting–Gathering
		9–12,000	Paleo-Indian	Early Hunting
Quartenary			*Continental Glaciation*	*Associated Deposits*
		12–20,000	Post-Wisconsin	
		20–150,000	Wisconsin 3d Interglacial	Alluvium, loess (eastern Nebraska), & dune sands (Sandhills)
		400–550,000	Illinoian	
	Pleistocene		2d Interglacial	
		1–1.4 million	Kansan 1st Interglacial	Sand to coarse gravel alluvium (eastern Nebraska & Platte Valley)
		1.7–2 million	Nebraskan	
			Geologic Stratum	*Associated Localities*
	Pliocene	2 million		
		5 million		Not represented in Nebraska
		6 million		
	Miocene		Ogallala Group	Poison Ivy quarry Ash Hollow cave Scott's Bluff (top)
Tertiary		18 million		
		24 million	Arikaree Group	Agate Fossil Beds
	Oligocene	28 million		
			White River Group	Scott's Bluff (base)
		35 million		
		38 million		
	Eocene			Not represented in Nebraska
		54 million		
	Paleocene			
		65 million		

southwestern borders to the northeastern corner of the state and south across the central Platte Valley.

The animals of the earlier parts of the Cenozoic era in Nebraska include such strange creatures as the enormous titanotheres, which survived from Eocene times to about 30 million years ago, or early Oligocene times. Another major group were the oreodonts, piglike creatures that were among the most abundant and widespread animals of the Oligocene and which survived into the middle Pliocene. Also in early Oligocene times many large predatory mammals were appearing, including both biting and stabbing cats and ancestors of coyotes and wolves.

During Miocene times, which began about 24 million years ago, the climate turned cooler and drier, favoring grassland species over forest-adapted ones, and a host of large grazing animals appeared in Nebraska. Fossil beds some 21 million years old near Agate Springs suggest that among the most numerous of these animals were horse-like chalcotheres and various rhinos. The rhinos there were part of a lineage that had begun in the Oligocene and that continued through the late Miocene. There were also giant pigs, primitive beavers, and four-horned relatives of the antelopes.

Toward the end of the Miocene period, about nine million years ago, volcanic activity in the Rocky Mountains (and elsewhere) continued periodically. Thus, areas such as the Yellowstone Plateau continued to be shaped and molded. One such volcanic eruption, from an unknown western source, was sufficiently great to carry enormous clouds of volcanic dust into Nebraska and deposit it haphazardly over various parts of the region. For example, in the Niobrara Valley a deep mantle of volcanic dust suddenly descended on a marshy area, trapping and killing in place dozens of horses, rhinoceroses, and other animals that had been feeding there. In this way, a fossilized "snapshot" of the late Miocene life was preserved until it was recently discovered. This area, called the Poison Ivy quarry, gives us an insight into the life of the Great Plains nine million years ago, just as the excavations of Pompeii, buried by an eruption of Mount Vesuvius, show us a glimpse of the Roman Empire in 79 A.D.

By the end of the Pliocene, about two million years ago, Nebraska was probably an undulating plain, which scattered rivers wandering through it and vast herds of grazing mammals foraging on its lush vegetation. Fossil remains show that ground sloths, rabbits, various kinds of rodents, primitive horses, wild peccaries, camels, llamas, and four-horned prongbuck antelopes were present.

During the early stages of the Pleistocene, the Nebraskan and Kansan glaciations deposited great quantities of glacial till over the eastern quarter of Nebraska and no doubt greatly changed the ecology of the region. Now a variety of elephants (including both mammoths and mastodonts) and other arctic-adapted forms such as muskoxen entered the area. (The woolly mammoth probably never actually extended its range as far south as Nebraska.) Many of these mammals had reached North America from Asia during major glacial periods, when vast amounts of water were converted to ice and snow, lowering ocean levels several hundred feet and thereby connecting North America and Asia with a land bridge in what is now the Bering Sea. Thus the first mammoths reached this area about one and a half million years ago, and these early arrivals were about the size of modern Asian elephants. Later, considerably larger forms evolved, reaching their maximum size at the time of the last major glaciation. These large grazing animals sometimes fell prey to saber-toothed cats, or other leopard or jaguarlike cats. Giant bears, wolves, and early forms of coyotes also shared in the feasts provided by these animals. It was probably the greatest assemblage of gigantic and diverse mammals ever to have occurred in Nebraska.

During the latter part of the Pleistocene, at the time of the Illinoian glaciation, the ice sheets reached only northeastern Nebraska. Although many of the earlier large mammals persisted, the last of the ground sloths then disappeared from Nebraska, as did the giant bears and the American mastodonts. Disappearing too were the great mammoths, and the earlier giant bison was replaced with a smaller form, *Bison antiquus,* which closely resembled the modern species. Other fairly familiar creatures were also appearing, such as deer, elk, badgers, red foxes, and various mice and other small rodents.

Ice of the most recent Wisconsin glaciation, which terminated less than twenty-five thousand years ago, did not enter the state but affected Nebraska in a similar fashion. The fossil evidence for mammal life in this period is rather scanty. However, it is known that tundra-adapted forms such as muskoxen, caribou, and lemmings ranged as far south as Nebraska at that time. By then, vast amounts of water were flowing down the Missouri drainage as outflows from the melting glaciers to the north, and likewise the Platte River was no doubt draining similar glacial meltwaters in the Front Range of the Rockies. With the final retreat of the Wisconsin ice sheet the last mammoths also disappeared from Nebraska. So too did *Bison antiquus,* which was replaced by the modern American bison. The last of the native American horses also died out, and horses were not to appear again in North America until they were brought in by Spaniards during historic times.

At the final stages of the Wisconsin glacial period, or about twenty thousand years ago, it is likely that a new creature set foot on the North American continent. This most probably happened sometime over thirteen thousand years ago, just before the Bering land bridge disappeared for the last time. Thus the bridge that had allowed mammoths, muskoxen, and other large mammals to cross into North America also offered an avenue of immigration for the creature that was ultimately to reshape the continent itself. The animal was a strangely bipedal primate, later to be known as man.

After Tirawa had created the sun, moon, stars, the heavens, the earth, and all things upon the earth, he spoke, and at the sound of his voice a woman appeared upon the earth. Tirawa spoke to the gods in heaven and asked them what he should do to make the woman happy and that she might give increase. The Moon spoke and said, "All things that you have made, you have made in pairs, as the Heavens and the Earth, the Sun and the Moon. Give a mate to the woman so that the pair may live together and help one another in life." Tirawa made a man and sent him to the woman; then he said: "Now I will speak to both of you. I give you the earth. You shall call the earth 'mother.' The heavens you shall call 'father.' I give you the sun to give you light. The moon will also give you light. The earth I give you, and you are to call her 'mother,' for she gives birth to all things."

Pawnee creation story
(Dorsey, 1906)

The First People

The first people to see the Platte left no enduring footprints. Most probably they were a band of Paleo-Indian hunters, following herds of mammoths about 10,000 B.C. Archaeological data indicate that then a small group of mammoths was killed and butchered along the South Platte River, near the present location of Dent, Colorado. Among the skeletal remains of at least twelve mammoths were two spear points of a type known as Clovis Fluted. These spear or lance points, which range in size from less than two inches to about five inches long, and with a width less than half their length, have broad and distinctive shallow grooves up one or both sides for about half their length.

With such simple weapons the early elephant hunters of the plains were seemingly very poorly equipped to deal with the Columbian mammoth, an awesome beast that stood thirteen or fourteen feet tall at the shoulders and had tremendous tusks that were twice as long as those of any modern elephant. Without horses or other animals with which to hunt from, these enterprising people must have concentrated on killing the weaker or injured animals, or perhaps hoped to come on a group trapped in a water hole or quicksand. Unlike the later Indians, who drove bison over cliffs or caused them to stampede by using prairie fires or horses, the earliest Indians of the plains must have found the mammoths to be extremely dangerous and uncertain prey. Furthermore, the mammoths and mastodonts were nearly at the end of their long period of prominence on the North American scene, and by 8,000 B.C. they were nearly extinct, perhaps as a result of climatic changes. Several other large mammals, including giant bison of the *latifrons* type, also disappeared at about the same time, perhaps in part because of hunting pressures from early man.

The glaciers retreated and giant bison disappeared from the southern and central plains during late Wisconsin times. The bison were replaced by *Bison antiquus*. Although smaller than *latifrons, antiquus* was still somewhat larger than the present-day bison and had longer and straighter horns. By 6,500 B.C., great herds of these creatures were wandering across the entire plains region; they were hunted by prehistoric Indians in an area from the Prairie Provinces of Canada to Texas and New Mexico and from the Rockies into the Dakotas and western Nebraska.

These early buffalo hunters were equipped with stone points that were smaller and more beautifully made than the Clovis points. A broad groove on one or both faces runs most of the way to the tip and produces two ridges parallel with the edges, which are rounded or tapered. The base often has two small projections toward the rear, and sometimes a small protuberance at the center. Altogether, it is a much finer piece of workmanship than are the earlier points, and like them it probably served as a spearhead or lancehead. Much later, stone points known as Plainview tips were made; these are somewhat similar to the Clovis type, but lack fluting.

On Lime Creek, a tributary of the Republican River, an archeological site of the Paleo-Indian tradition has been found that, in its older layers, was probably active about ninety-five hundred years ago. Here, a variety of jasper blades (up to seven inches long), hammerstones, end scrapers, and parts of a shaft smoother have been found. There were also the bones of seventeen kinds of mammals, including beavers, various birds, and reptiles. These people were evidently quite opportunistic hunters. In a nearby site, with estimated dates ranging from slightly more than five thousand years ago to more than ten thousand years, a large number of animal bones were found, including those of bison, antelope, deer, coyotes, rabbits, mice, rats, and prairie dogs. There were also the bones of birds, reptiles, and amphibians and burned nests of mud-dauber wasps, whose larvae were probably eaten. Many leaf-shaped stone points were present, as well as scrapers, blades, drills, grinding stones, bone needles and awls, and other bone or stone tools. From these findings it is clear that the earliest human hunters of the Platte and Republican valleys probably were small groups of people who relied not only on bison for their meat but killed whatever was available. They probably killed the bison by stalking them carefully, perhaps under a wolf skin, and bravely speared them or lanced them to death. Or fires may have been laid in the paths of grazing bison, causing them to stampede and plunge over ridgetops or into deep creekbeds or water holes, where the injured animals could be more safely dispatched.

During these years, the climate slowly warmed. So, from about seven thousand years ago until about twenty-five hundred years ago the grasslands of the plains periodically dried out under the summer sun, only to be regenerated by winter snows and spring rains. On these high plains, a new and somewhat smaller species of bison gradually replaced the earlier type about five thousand years ago, and the Indians of the area adapted to this source of food. At Signal Butte, in the North Platte Valley, a group of people lived some three thousand to thirty-five hundred years ago. While probably primarily buffalo hunters, they also probably gathered various vegetable materials and ground them for food. They used hearths, and they dug small pits that were evidently used for storage. Materials like charcoal and burned stones and various other debris associated with their sites suggest that fire was important to them. Bison bones discovered here are of the modern species, and a variety of stone and bone tools and projectile points have also been found at their sites.

A somewhat older site in Burt County at Logan Creek, a tributary of the Elkhorn in the lower Platte drainage, had an abundance of animal bones, especially of bison. There was also present a type of projectile point similar to that of the Eastern Woodland culture, in which notches were placed on the sides near the base. Thus, the woodland-adapted Indians had carried their influence at least this far west by then. Similarly, around 4000 B.C., bison hunters were living in the Red Willow Valley, a Republican River tributary located near the present site of McCook, Nebraska. Excavations of sites here have unearthed projectile points similar to those of the Logan Creek site, and apparently indicate a westward extension of this culture well into the high plains. Much later, between 140 B.C. and 780, the same area was occupied by Plains-Woodland peoples, and at the peak population period, around 1200–1500, a group of hamlets that subsisted on a mixture of hunting and maize growing were present in the valley. It is not certain when the development of a maize-growing culture first appeared in the Platte Valley, but an early Hopewell site found near Kansas City, in which maize was clearly present, dates from the first few centuries of the Christian Period, or about seventeen hundred years ago. Most probably such native plants as sunflowers, pigweed, lamb's-quarters, and similar seed-bearing flowers were planted and harvested even earlier.

Well before two thousand years ago a Plains-Woodland culture had been flourishing throughout the eastern plains, and it extended west along the rivers and creeks of Kansas and Nebraska. These people produced a type of pottery for food storage that was heavy and coarse, with wide openings and sometimes with simple decorative effects produced by poking the clay with a round-edged stylus. At least some of their sites in Nebraska suggest that they constructed rather simple thatched structures, perhaps similar to those of the agriculturally based Eastern Woodland peoples. Small pits were dug near the structures for food storage, but the presence of awls and needles for working with skins, as well as various fleshing tools, suggest that animal hunting was probably an important means of survival. Furthermore, objects made from the bones of deer, antelope, and bison suggest a diversity of prey. The earliest of these Woodland type-sites in

Nebraska has been dated around 2000 B.C., and the most recent ones reach well into the Christian Period.

However, at some time at least as early as 900, and perhaps as early as 500, these peoples were replaced with a rather different culture, whose sites are marked by definite evidence of cultivation as well as hunting and fishing. In Nebraska these sites are located on the Republican and lower Loup rivers, thus encompassing the Pawnee homeland in the Platte Valley. These people lived in daub and wattle houses of substantial size. The dwellings were square or rectangular, with long covered entrances that usually faced east, away from the prevailing winter winds. The houses were grouped in small villages, usually along bluffs or ridgetops, and sometimes also on lower river terraces. The villages were always located near riverbottom land that could be planted to maize and other crops, and they had storage pits for protecting their harvests. Inhabitants of these sites fished in the nearby streams and creeks and gathered wild fruits and berries in season. Hunting was done by bow and arrow, and domesticated dogs enlivened the camp scene. Tobacco was smoked in stone or clay pipes, and various kinds of ornaments made from bone and shell were also used. The pottery was simple, typically consisting of round-bottomed jars with thickened rims and uncomplicated decorations on their exteriors.

Near the confluence of the Platte and Lower Loup rivers, especially on the north bank of the Loup, are a group of at least a dozen village sites of this general age (1600–1750) and culture type. Known as the Lower Loup or early Pawnee type, they consist of circular earthlodges up to fifty feet in diameter. A typical earthlodge contained a depressed floor and a central fireplace; the roof was supported by four main poles, and a series of smaller posts supported the sloping roof. A long, covered entrance usually faced toward the east, and cache pits were dug into the floor of the dwelling. Each of these large buildings probably housed up to twenty persons, and the abundance of the cache pits suggests that crop growing was well developed. Hunting with bow and arrow certainly was also important, and large dogs were probably used as beasts of burden for carrying or dragging equipment from place to place.

At approximately the same time, another cultural unit, which relied primarily on hunting, was living along the small creeks (such as the Dismal River) of the Sandhills and the Platte and Republican rivers. They did not live in earthlodges, but instead probably constructed small structures that may have been covered with skins, grass, or thatch; and they rarely stored food in cache-pits. Probably few crops could be grown in these sandy or dry sites, and the people must have survived largely on hunting, gathering plant materials such as berries or roots, and the like. The Dismal River culture is dated from 1650 to 1750, or at the time when the Spaniards were moving north into the southern plains and the French explorers were reaching west into the Missouri Valley. It has been suggested that the Dismal River people represented the northernmost remnants of an early southern Apache cultural invasion into the central plains, and they have been called the Lipan Apaches.

The origins of the Pawnees are shrouded in the past. Their territory once spread from the Niobrara River in the north to the Arkansas or Canadian River in the south, and from the Missouri River westward an unknown distance toward the Rocky Mountains. However, the west and east boundaries were severely limited because of conflicts, initially with the Comanches, later with the Teton Sioux to the west, and still later with the Omahas and Otos to the east. At the Pawnees' southeastern limits they were met by the Kansa Indians, who dominated the region from the Big Nemaha River south to the Kansas River and west to the Republican River, and to the southwest they encountered the hunting grounds of the Apaches and Kiowas. Thus, the Pawnees were essentially confined to the present limits of the state of Nebraska and north-central Kansas, with the land to the south largely used for wintering and hunting grounds. Their major villages were all associated with larger river valleys, and all of the most sacred places of the Pawnees were located near rivers or springs. There were five of these sacred places. One was on the south bank of the Platte opposite the present site of Fremont; another was on an island in the Platte near Central City; one was on the Loup Fork near Cedar River; one was on the Solomon River in Kansas; and one was along a spring near Guide Rock, Kansas.

Although the Pawnees may well have developed in place from their earlier Upper Republican River culture, it has also been suggested that they may have moved into Nebraska from the south. According to one account, they may have come to Nebraska from a homeland along the Arkansas River of Oklahoma. One of the earliest of these groups to reach Nebraska was supposedly a group called the Kawarahkis, which lived for a time at the confluence of the Little Nemaha and the Missouri rivers before colonizing the areas of the Platte and Republican rivers. The Pawnees eventually colonized virtually every stream and river in eastern Nebraska, but were centered in the more northerly areas. One group, the Grand Pawnees, were located on the south bank of the Platte. An offshoot of this group initially moved into the Republican River area, but early in the 1800s they moved north to join the other bands on the Platte and Loup rivers, probably because of harassment from the Kansas to the south. At the time of their greatest extent, the Pawnees probably controlled much of the area along the Platte, Loup, Republican, Blue, and Smoky Hill rivers. Yet, most of these were gradually given up, and by the early 1800s the Pawnees were largely confined to the Platte and Loup valleys.

The Skidi Pawnees were often called the "wolf Pawnees" by other groups (the Loup River, home of the Skidis, means *wolf* in French). Indeed, *Skidi* is perhaps only a transliteration of *Skiri-ki,* meaning *wolf,* and refers to their great endurance as well as their uncanny ability to mimic wolves by placing a wolf-skin robe over their backs and advancing on hands and knees in the manner of a wolf. In this way the Skidis could easily approach bison or possibly get close enough to enemy camps to steal some of their horses. When hunting bison, several men would approach a small herd of bison from various directions, until the group was completely encircled; then, panicking the animals, they closed in on them with bows and arrows, killing and wounding as many of the beasts as possible in the confusion.

Besides being good hunters, the Pawnees were fine farmers. They planted not only corn but also beans, pumpkins, and squashes, and they dug up the roots of many edible plants. Indeed, until the central plains cultures obtained horses, which increased their capabilities for hunting and reduced dependence on corn, bison were probably only a secondary means of subsistence for them, for the corn plant was a sacred mother entity, *a-ti'-ra*. The Pawnees also had a special love of birds and their feathers, and quite possibly the name *Pawnee* is derived from this fact. Thus, the culture was known as the *Pani* by the French, *Panana* by the Spanish, and *Panani* by the Sioux. This last variant is also used in the name of the hairy woodpecker (*zong-panani*), suggesting that it may have referred either to the birds' excavation abilities, or more probably to its colorful red crest. It seems likely the latter explanation is the better one—the Pawnee tradition of wearing red feather ornaments or red crests may well have been the basis for calling themselves the *Panyis*. Further, the Caddoan culture to the south, from which the Pawnees were derived, used the term *banit* for a kind of migratory bird that served as an important omen. This omen was carried on into Pawnee life as the Hako ceremony, a bird-oriented ritual lasting for several days and associated with peace. Alternatively, it has been suggested that the term is simply an abbreviated version of *Pariki,* meaning a horn and referring to an erect scalp-lock.

The Pawnees knew the Platte River very well; they called it the *Kizkatuz,* referring to flat or shallow water, or the *Kisparuksti,* meaning "wonderful river." Near one of their major early villages on the Platte, close to the present site of Clarks, was an island home for the Pawnees' animal spirits, or *nahuarac*. This island was *Lalawakohtito,* the "dark island." From the forks of the North and South Platte rivers eastward almost to the mouth of the Elkhorn, many campgrounds or villages of the Pawnees were located, including at least one major village, and they had many winter hunting camps farther west along the Platte. The group living along the Platte was known as the Chau-i ("Grand") group, to distinguish them from the Skidis to the north and the others living farther south on the Republican River. To the early French traders they were known as the Grand Pani, apparently because of their tall stature and long strides. Pawnees living on the Republican River were known as the Pita-hau-erat ("down the stream") and the river was named thus by the French because of the Indians' independent, republiclike nature. The Grand Pawnees and those of the Republican River area tended to do their winter hunting in the southern areas, on the Republican and other tributaries of the Kansas, while the Skidis dominated the areas to the north and west of the Platte all the way to its fork.

An intrusion of the Grand Pawnees into Skidi hunting areas north of the Platte in the late 1700s brought an extended period of hostilities between these two groups, and it eventually resulted in the defeat of the Skidis and their acceptance of the Grand Pawnees as leaders of the entire group. Yet the Skidis refused to be totally dominated, and well into the mid-1800s they continued to fight periodically with the Grand Pawnees. But by then intertribal problems were secondary to those being caused by the whites.

I [Tirawa] will show you how to build a lodge, so that you will not be cold or get wet from the rain. Go and get timber. Cut ten forked sticks and set them in a circle. Four of the upright forks must form a rectangle, with the longest sides extending east and west. The posts that are set in the ground to uphold the lodge represent the four gods who hold up the heavens in the northeast, northwest, southwest, and southeast. There are minor gods between these, with powers that connect the power of one god to another. There is also an outer circle of many gods, and you shall cut poles to represent them; their power also extends from one god to another.

Pawnee creation story (Dorsey, 1906)

A view of the Platte in early spring, showing plant succession on islands and sandbars. Photo by William S. Whiteny, Flatland Impressions, Aurora, Nebraska

Behold on Mother Earth the running streams!
Behold the promise of her fruitfulness!
Truly, her power gives she us.
Give thanks to Mother Earth who lieth here.

Pawnee song (Densmore, 1929)

The River

The river was born before man ever looked upon the plains; its age is impossible to guess. Perhaps its antecedents go back to more than fifty million years ago, when the Rocky Mountains were rising and beginning to intercept moisture-laden winds. As the rain and snows that had pelted the crests of the infant mountains gathered into rivulets and streams, some of these must certainly have tumbled down their easterly-facing slopes and tentatively wandered out onto the drier plains, where they began to crease a watery trail across the more gently sloping lands toward the east.

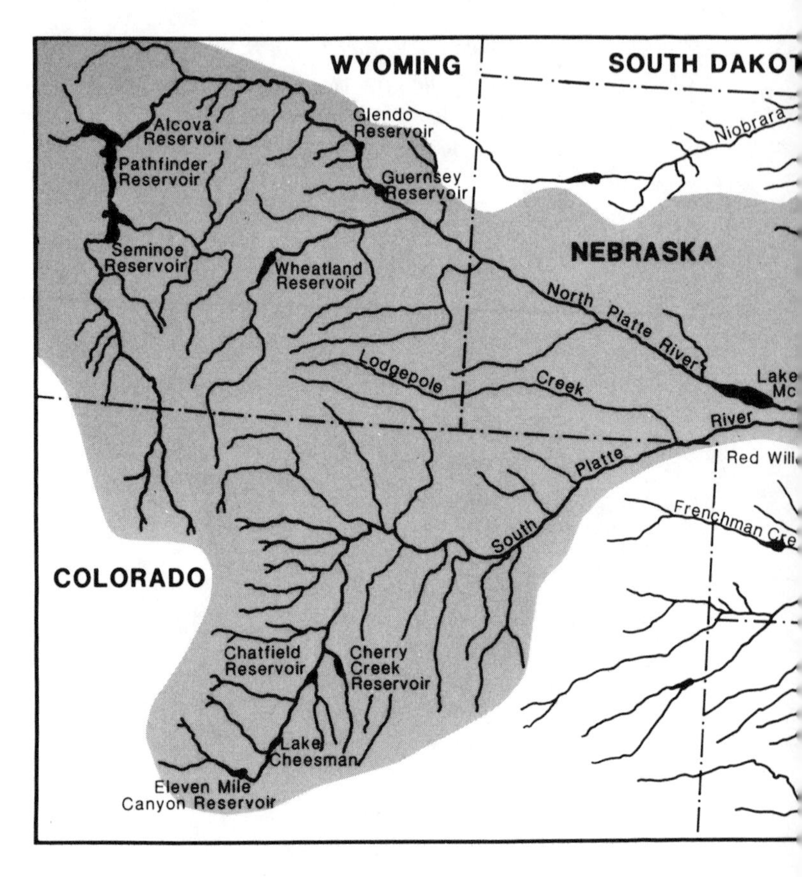

FIGURE 1: Map of the Platte River basin (shaded), exclusive of the headwaters of the North Platte tributaries in the Sweetwater drainage. The Niobrara River basin and Kansas River basin are also indicated.

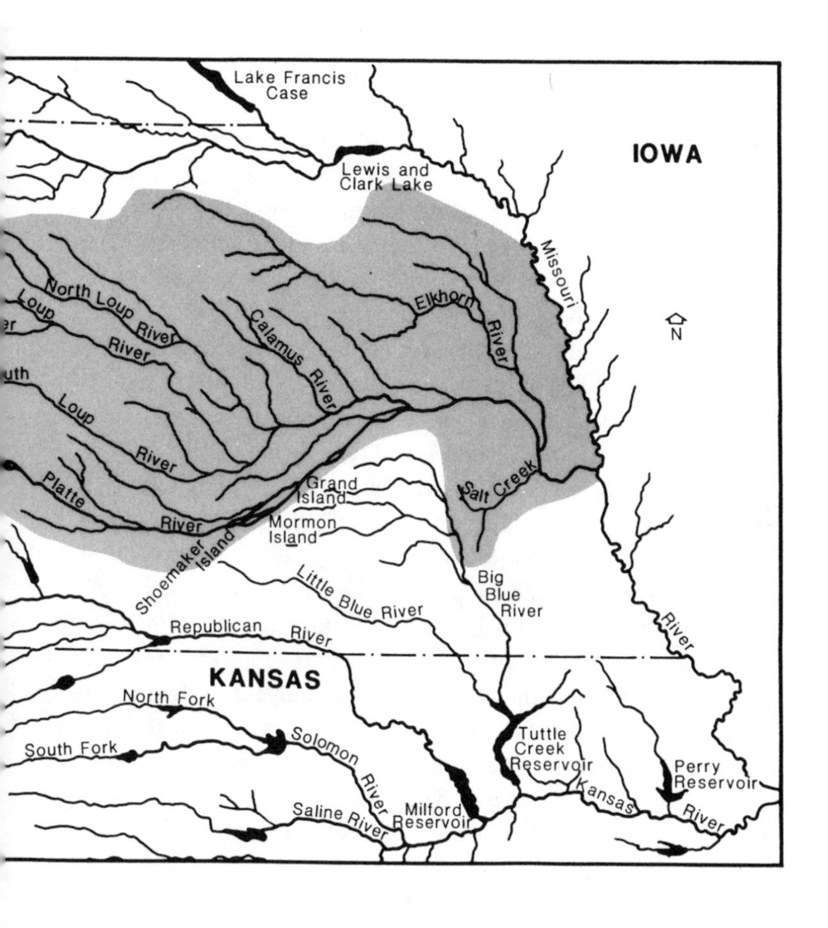

Lake Francis Case
Lewis and Clark Lake
IOWA
Missouri
N
North Loup River
Loup River
Calamus River
Elkhorn
River
Loup
River
Salt Creek
Platte
River
Grand Island
Mormon Island
Shoemaker Island
Little Blue River
Big Blue River
River
Republican
River
KANSAS
North Fork
Solomon
River
South Fork
Tuttle Creek Reservoir
Perry Reservoir
Kansas
River
Saline River
Milford Reservoir

By the end of the Oligocene epoch, or about 24 million years ago, a vast sheet of sediments had covered nearly all of western Nebraska. Much of this deposit was wind-carried silt, and it was rich in volcanic ash that had been generated by the mountain-building activity farther west. Still later, in Miocene times, deposits known as the Arikaree and Hemingford strata were laid over the Oligocene materials, and these in turn were buried by still later deposits of the Ogallala group. These deposits, laid down by eastward-moving streams, have mixed within them the seeds of grasses and herbs that suggest that by this time the surrounding lands supported a grassland and meadow vegetation. Two of these streams were eventually to coalesce and become the ancestral Platte River.

In early Pleistocene times, a series of glaciers began to have their effects on the life and land of the central plains. This glacial period, the Nebraskan, brought with it outwash deposits of the glacier itself to eastern Nebraska, and probably also at the same time gravelly sediments were carried into the state by the infant North Platte River, which was then being nourished by the increasing runoff from the western mountains. Although the general drainage system of the Platte had been established by the end of the Pliocene period, it was now greatly affected by the combination of the materials carried down by ice sheets in eastern Nebraska and the increased load of materials carried in from the western headwaters of the river. As a result, channels repeatedly changed and the river meandered over much of the state. At one time in the early Pleistocene, uplifting of deeper rock layers in western Nebraska was even sufficient to deflect the early Platte into a drainage that carried it from the present North Platte channel northeastward into the present Niobrara drainage. Somewhat later, during the peak of the Kansan glaciation, massive ice sheets moved through eastern Nebraska, bringing down enormous amounts of glacial till and again diverting the course of the stream. This time it was sent south, into what is now the Republican River drainage. During this same period, sediments of the Grand Island Formation were widely deposited over central Nebraska. These sediments were mostly sands and gravels associated with the inwash and outwash of the glacial till.

During the third period of glaciation, the Illinoian, the Platte River finally settled into its present course and began to erode a series of channels into the sands and gravels of the recently deposited Grand Island sediments. During this period, greenish clays and sands of the Illinoian glaciation were being deposited in eastern portions of the state. Still later, during the Iowan glacial phase, the lower Platte Valley continued to erode, while during the same time deposits of reddish sand, gravel, and wind-carried dust came into the state from elsewhere, perhaps from the west or south.

Finally, during the last major glacial period, the Wisconsin, wind-deposited materials were dropped widely over the eastern part of Nebraska, and at the same time the Sandhills to the north of the river were being formed. As strong west winds shifted these sandy deposits about, the lighter particles were carried east and south by the wind, where they were to form an extensive deep mantle of fine loess soil capping the earlier reddish and greenish sands, gravels, and wind-carried materials of the Iowan and Illinoian glaciations.

The river must have been greatly affected by the massive reshaping of the land, and the areas through which it doggedly wound must have changed enormously over the eons. By the early stages of the Wisconsin glaciation, however, the river was probably passing through a grassland-dominated land not very different from the present-day scene. However, both to the north and south of Nebraska it is probable that a varied mixture of coniferous and grassland vegetation types may have stretched from the Dakotas south to Texas. Here, one might imagine, caribou and mammoths grazed on the sweet grasses of the prairies, and bison wallowed in the muddy pools. At the time of the maximum advance of the Wisconsin ice sheet, only some twenty thousand years ago, it is probable that conifers and other subarctic plant species also grew abundantly along shaded river valleys like the Platte's, while in warmer localities deciduous forests probably maintained local control. In somewhat drier sites the prairie grasses asserted their dominance.

The Wisconsin glaciation did not end easily, and between seventeen thousand and ten thousand years ago there were occasional periodic readvances of the ice and associated alternations of warmer and cooler climates. Between nine thousand and four thousand years ago there developed a rather protracted period of warm and relatively moist conditions, which perhaps allowed some elements of the eastern deciduous forest to spread westward along the Platte's edges, and probably also forced a gradual elimination of coniferous trees from some of these same areas.

Still more recently, in the period between four thousand and twenty-five hundred years ago, the climate continued to warm but also became drier, causing extinction of all but the most tolerant of the northern species. This period also allowed the grasses to spread inexorably outward to do battle with the intruding forests, and to become victorious in all but the most moist and shady river valleys. Withal, the river forever flowed onward, always bringing new water from the Wyoming and Colorado mountains, nourishing the woody plants that huddled ever more closely to its shorelines, and departing as quietly from the state as it had arrived. Its water gave life, and it took only the silt or sand it carved out of its twisting shoreline. Even this slight tribute was given up as easily as it had been taken, for the river was prone to deposit materials wherever its course was impeded. Thus, new shorelines were bestowed on the land just as frequently as they had been borrowed from farther upstream. So, new bends, channels, and isolated oxbows were forever being formed and reformed, molding and recasting the landscape over its broad and infinitely pliable flood plain. The river was to become the lifestream of the entire area, and the trails of countless herds of bison were cut into the land and radiated out from the river's edges like branching blood vessels, as the animals annually made their migrations up and down the central plains of North America.

"Listen," he said,
"There the buffalo are coming in a great herd,"
These are his sayings.
"There the buffalo are coming in a great herd,
"The waves of dust roll downward,
"There the buffalo are coming in a great herd,
"They mark the place of the buffalo wallow."

Pawnee song (Densmore, 1929)

From the sandhills [the Platte] had the appearance of a great inland sea. It looked wider than the Mississippi and showed to much better advantage, there being no timber on the banks to check the scope of the human eye. Grand Island which lays just opposite in the middle of the river is one hundred miles long, and has some cottonwood trees upon it. . . . The water is exceedingly muddy, or I should say sandy; and what adds to the singular appearance of this river, the water is so completely filled with glittering particles of micah or isingglass that its shining waves look to be rich with floating gold.

Immigrant diary, 1850
(Mattes, 1969)

The Next People

The first white men to have left a permanent record of the Platte River were French explorers, the Mallet brothers, Paul and Pierre. In June of 1739 they named it *La Rivière Plate,* "flat river." Although earlier fur traders or missionaries had no doubt crossed the mouth of the Platte on their travels up the Missouri River, during that time it seems to have been known only as the *Rivière des Panis,* "River of the Pawnees." Thus, maps dating from as early as the 1690s indicate the locations of Pawnee villages along the tributaries of the Missouri. The Mallet brothers reported that they followed the Platte's course for some twenty-eight leagues (or seventy-five miles) upstream, to the mouth of the "River of the Comanches." This was most probably the point of confluence of the North and South Platte rivers, and the original limits of Comanche hunting territory.

There is other evidence that the Mallets were not the first whites ever to visit Pawnee lands, for a Spanish military expedition led by Pedro de Villasur may have reached the vicinity of the Platte River as early as 1720. Certainly by the late 1700s the Pawnees were regularly encountering and trading with French fur traders who were moving westward up the Platte River. Furthermore, there was a trade in British goods in the areas of the Otos and Pawnees, an activity aided in part by the Omaha tribe, which helped to win the Otos over to British interests. But Spanish influence also was felt, and the Pawnee chiefs occasionally received Spanish medals as gifts of the Spanish crown.

At the time of these early trading contacts there were Grand Pawnee villages on the south bank of the Platte near the present site of Linwood and southeast of the present location of Clarks, in Polk County. There was also a large village of Skidis on the north bank of the Loup. An early missionary, John Dunbar, estimated that at the time of his arrival in the area in the mid-1830s there might have been as many as eight thousand to ten thousand Pawnees in the Platte and Loup River villages, but their ranks were soon to be thinned by disease and warfare.

The first exploration of the Platte by an American party occurred in 1820, when Lieutenant Stephen H. Long led an expedition up the Platte River as far as the South Platte. He referred to the area as the "Great American Desert" and considered it to be uninhabitable. Yet the first wagon train from St. Louis went up the Platte Valley in 1830, the same year a party from the American Fur Company left Bellevue and traversed westward along the north bank of the Platte. Six years later the first white women, the wives of missionaries, accompanied a group of fur traders heading west from Bellevue.

In 1841 the first party of immigrants, numbering about one hundred, left the Kansas River area to follow the Oregon Trail west along the Platte Valley. Two years later they were followed by a group of nearly 1,000 people who departed from Independence, Missouri. This group included 130 women and more than 600 children, all bound for Oregon to try to help ensure its future as a part of the United States rather than the British Empire. In 1845 several thousand more people left for Oregon, and during that year Colonel Stephen Watts Kearny responded to a call by President James Polk to establish a series of forts along the route to Oregon. He judged that during the season of 1845 approximately 850 men, 475 women, 1,000 children, 7,000 cattle, 400 horses and mules, and 460 wagons had traversed the Oregon Trail. Fort Kearny was established on the south side of the Platte west of Grand Island in 1847. It was initially called "New Fort Kearny," to distinguish it from an earlier post that had been established at Table Creek (Nebraska City) and later abandoned.

By 1848 the fort was little more than a crude group of log and sod huts and a shabby-looking attachment of soldiers. It had been situated far enough west on the Platte to receive migrants who had begun their trips from a variety of starting points. Some came from St. Joe or Independence, Missouri, and had followed the Little Blue River northwest until they reached the Platte. Others had come directly west from Nebraska City, crossing dry country until they reached the south shore of the Platte at various points. Still others headed west from Plattsmouth, or from various jumping-off places in the Bellevue, Omaha, or Council Bluffs areas. These latter departure points followed the north bank of the Platte River, and such was the route taken by Brigham Young and his Mormons on their way to Great Salt Lake in 1847. Thousands more Mormons followed in 1848, and by 1849 the massive migration to California began. Perhaps as many as thirty thousand people passed along the Platte Valley in 1849, and nearly sixty thousand followed the next year.

The year 1850 was a watershed year for the Platte Valley emigration, for although a record number of people passed through the Platte Valley, it was also a year marked by a cholera epidemic and drought that killed both people and animals by the thousands. Yet the hordes of emigrants continued, numbering some fifty thousand in 1852 and perhaps twenty thousand in 1853. Of the possibly one-third of a million people who traveled west along the Platte between 1841 and 1866, almost half passed through during the five-year period from 1849 to 1853. The end of this era was not to come until 1869, with the completion of the transcontinental railroad system that linked the coasts and made the Oregon Trail permanently obsolete.

Besides leaving hundreds of graves of emigrants who had died of cholera, other illnesses, misdeeds, or sheer bad luck, the emigrant trail also left behind a scattering of settlers who decided for one reason or another not to follow their dreams all the way to California or Oregon, but instead to settle for a life in the Platte Valley. Thus, near Fort Kearny, "Dobytown" arose, named for its "adobe" or sod houses. Later, the present town of Kearney grew up across the river. Fifty miles downstream a group of German immigrants from Iowa decided that they would establish a town on lands that had been recently ceded by the Pawnees. They hoped in vain that it might perhaps later be chosen as the site for a new national capitol because of its central position in the continent, and considered calling it "New Philadelphia." Such a fate was not to develop for Grand Island, but the little village along the Platte River would eventually become the third largest city in Nebraska.

The original settlement of Grand Island closely hugged the bank of the north channel of the Platte. The settlement was initiated in 1857 by thirty-five persons who had followed a small advance party of surveyors that had left Davenport, Iowa, in early spring. They placed their settlement near the lower end of a sixty-mile-long island that had been formed when a small channel of the Platte had separated and then rejoined the main channel, enclosing an area of land known to the French fur traders as the *Ile de Grande*. A second contingent of German settlers arrived the following year; most of this group spread out in various directions from the original settlement. In the fall of 1858 the town was badly damaged by a prairie fire that destroyed many of the newly built homes, and in 1862 the first of a series of migratory locust plagues appeared. Trouble with the Pawnees and other Indian tribes also began in the 1860s and continued on into the 1870s, when the locust populations also reached their peak.

As the Union Pacific Railroad construction period began in the Grand Island area in 1864, all of the larger trees in the area, which were mostly confined to the river islands, were quickly harvested. At about that time too an early settler named Jesse Shoemaker, who had moved into the Grand Island area from what is now Merrick County, discovered that these same islands made excellent summer pasture range for cattle. He took out a claim on an island that was later to become known as Shoemaker's Island, and successfully raised cattle there until the 1870s. But in November of 1871 a three-day blizzard descended on central Nebraska and caused great damage. Fortunately, Shoemaker's cattle had already been removed from the island by then, but only two years later they were caught in a blizzard nearly as bad. This storm, which occurred in April of 1873, drove many of the cattle into the river where they drowned. Shoemaker eventually abandoned the island, but it was soon homesteaded by others, and its original name persisted.

Immediately downstream from Shoemaker's Island is Mormon Island, named for the early Mormon emigrants that passed through the area by the thousands. Directly south of the Grand Island settlement, it was one of the first of the islands to be crossed by bridges, and soon afterwards it was settled. Perhaps the first pioneer on this section of the Platte was Friedrich Langman, who had arrived with the original band of settlers in 1857. He journeyed south from the townsite and selected an eighty-acre piece of land near the south channel of the Platte. He married a young woman who was also of German stock, and together they raised a family of three children along the banks of the Platte. His wife drowned during a spring flood on the Platte in 1883, but Langman continued to live on his land until the early 1900s.

As the emigrant trains moved west beyond the Grand Island area, they encountered Fort Kearny. Here in late spring, when most of the trains passed through, the Platte could be forded. The river consisted of as many as ten channels of varying sizes, with the total river span perhaps three miles, and the widest channel nearly two miles across. Yet these channels were nearly always very shallow, and so the Fort Kearny area provided an opportunity for persons coming down the Mormon Trail on the north side of the Platte to cross over to Fort Kearny. Farther downstream the river was more likely to consist of a single enormous channel, and there are no records of emigrant crossings of the river at Grand Island or beyond.

It was also in the vicinity of Fort Kearny that the emigrants were most likely to obtain their first views of buffalo herds. Although there are historical records of bison occurring as far east in Nebraska as the Little Blue River and even the Missouri, by the time of the great emigrant movement their primary range was centered on the forks of the Platte, and large numbers might be seen anywhere between Grand Island and Scott's Bluff. The bison provided a source of fresh meat for the travelers; furthermore, great numbers were slaughtered simply for the sport it provided. Coyotes and wolves followed the herds of bison, killing and eating the sick and wounded or simply scavenging the carcasses left by the whites. They rarely threatened isolated persons who happened to stray too far from their camp without weapons, and there are no records of anyone being killed by prairie wolves. Thus it seems that their continuous presence and calling at night was more a psychological danger than a real one.

At least until the late 1850s, or well past the peak of the emigrant movement through the Platte Valley, the Indians caused only minor problems. Occasionally they would try to steal a horse or other livestock from the camps, but more often the war parties that were seen by the whites were simply those of the Sioux or Cheyennes, on their way to attack the peaceful Pawnees. However, in the 1860s the Indians turned on the whites, largely in retribution for treaty violations in the Dakotas and Wyoming and the many acts of violence that had been dealt them by the emigrants. The 1860s were thus a time of terror on the plains as the Cheyennes and Sioux began to retaliate, wreaking havoc on the settlers and emigrant caravans. Beyond the forks of the Platte were the regular hunting grounds of the Cheyennes and the Sioux, and the continuing flow of whites through the Platte Valley, with their senseless slaughter of the bison herds, was to spell disaster for the Indians.

In 1820 the herds in the area of the forks of the Platte were so vast that a soldier under the command of Major Stephen Long then wrote that there were "immense herds of bison, grazing in undisturbed possession, and obscuring, with the density of their numbers, the verdant plain." But by the 1860s there were few bison to be found in the eastern parts of the Platte Valley. As late as 1867 about half of a herd of four thousand animals was reported to have perished in quicksand while trying to cross the Platte, but by 1870 there were virtually none to be found north of the lower Platte. Perhaps the last small herd was killed by Indians in the Dismal River area during the winter of 1882–83. They were the last of the tens of millions of animals that had once trod the Nebraska plains, and their demise marked the end of an era in Great Plains history.

Beyond Fort Kearny, it was still 110 miles to the forks of the Platte, and this splitting of the streams marked the decision point for people traveling to Oregon or to the California gold fields. Many of the emigrant trains that were headed for Oregon by traveling along the south bank of the Platte continued to follow the South Platte for a distance of from a few to as many as 45 miles upstream before fording the river, then going north to intercept the North Platte and the Oregon Trail. Those that crossed the South Platte at the crossing farthest upstream traveled nearly directly north to reach the North Platte in the vicinity of Ash Hollow. Here a wooded canyon provided a freshwater spring, a supply of firewood, and a sense of oasis.

But Ash Hollow was also to be the last stopping place for some of the travelers, for many people died during the cholera epidemics of the late 1840s and early 1850s. Perhaps as many as five thousand people perished from cholera during the 1850 season alone, or nearly a tenth of the total emigrants who trekked the Platte River Road that year. The mouth of Ash Hollow was but one of many places where makeshift cemeteries gradually developed, but many people also were simply buried along the trail as soon as they had died. One of the people who died at Ash Hollow in 1849 was an eighteen-year-old bride named Rachel Pattison, who was "taken sick in the morning, died that night." But relatively few emigrants apparently died of cholera at Ash Hollow, perhaps because of the fresh water provided by the spring there. Yet across the river, on the north bank, cholera was widespread during 1850, and hastily dug graves were commonplace that year.

The Oregon Trail between Ash Hollow and Scott's Bluff passes such historic landmarks as Court House Rock and Chimney Rock, which are among the first of the great rocky signposts marking the Oregon Trail. These eroded outcrops of Tertiary sediments provide a hint of the enormous amounts of materials that were laid down in Oligocene times and then largely eroded away during the last twenty or so million years of time in the high plains. Many of the emigrants paused long enough to scratch their names in the soft claylike faces of these awesome monuments. At least thirty graves mark the area around Court House Rock, and perhaps as many also accumulated around the base of Chimney Rock.

From Chimney Rock west there is an increasing number of buttes, hills, and distinctive landforms. None of these is more impressive than Scott's Bluff, perhaps the most impressive natural monument of the Oregon Trail between the Missouri River and the Rocky Mountains. Although it was seen only from a distance by emigrants on the north side of the Platte, those on the south side were forced to choose between a longer but safer route through Ribidoux Pass or a shorter but much more hazardous trail through Mitchell's Pass. Ribidoux Pass was used primarily during the period up to about 1850. However, in 1851 the route through Mitchell's Pass became the principal one, apparently after enough work on it was done to barely permit wagon travel through it. Even then the route was narrow and tortuous. It was bounded on both sides by towering peaks, and was rife with ruts, deep mud, and sandy areas that could easily disable a wagon.

Beyond Scott's Bluff it was only another 150 miles of fairly easy travel to Fort Laramie, and at that point the Oregon Trail gradually began to leave the North Platte Valley to begin the long overland haul toward South Pass. Fort Laramie was the last stop on the Platte River Road, and those persons who had survived the cholera and other misfortunes of the plains now had additional and equally serious hazards to encounter before they were to reach their promised lands.

This is our experience crossing Platte River; the meanest of rivers—broad, shallow, fishless, snakeful, quicksand bars and muddy waters—the stage rumbles over the bottom like on a bed of rock; yet haste must be made to effect a crossing, else you disappear beneath its turbid waters, and your doom is certain.

Immigrant diary, 1862 (Mattes, 1969)

The timber that shall grow upon the earth you shall make use of in many ways. Some of the trees will have fruit upon them. Shrubs will grow from the ground and they will have berries upon them. All these things I [Tirawa] give you, and you shall eat of them. Never forget to call the earth "Mother," for you are to live upon her. You must love her, for you must walk upon her.

Pawnee creation story
(Dorsey, 1906)

The Island

Long before the island existed, there was the prairie. Ever resilient to the winds of time, it had endured both drought and fire and had successfully eliminated the hardwoods and conifers from all but their favorite retreats. Now it stood dominant over nearly all of the plains from the foothills of the Rockies to the Missouri River and beyond. The plants that constituted the prairie were older than much of the land they grew on. Evolving during the periodic drying and warming periods of the Tertiary, they had spread into the plains from more arid southwestern regions and had found a new way to cope with the bitter plains climate. They flowered and set their seed crop early, and then died back to ground level every fall, after the last bits of moisture from the previous winter's snows and spring rains had been wrung from the soil. They thus escaped both the effects of late summer fires and the killing cold of winter.

In the soft litter of the prairie soil, under an insulating blanket of snow, the prairie plant buds waited out every long winter, always ready to venture forth with the first warmth of the spring sun. Deep below, their roots extended down a dozen feet or more, extracting the subsoil moisture and deeper nutrients. Each summer the dying above-ground leafy parts added their contributions to the land, thus again returning their own nutrients to the soil and slowly enriching it.

Such was the cycle of life on the plains for millions of years. During this time hundreds of grasses and other prairie plants evolved to fill every ecological niche, surviving in habitats as diverse as sandy hilltops and rich, moist bottom lands. Some, such as big bluestem and Indiangrass, stood five to six feet tall, their rusty-red leaves and flower heads turning the fall prairie scene to the color of burnished copper. Other more modest grasses such as little bluestem and sideoats grama grew only about half as tall and formed less conspicuous ranks among the prairie assemblage. Still smaller were various other grama grasses and buffalo grass. These small plants survived best on the moisture-poor soils of the high plains, where they would flower and set seeds early in the summer months, long before the searing heat of August.

Wedded permanently to the soils that they themselves had formed, these prairie grasses reached from the tallest hilltops almost to the very shorelines of the Platte. Here, on wet, subirrigated meadowlands, they were among the first plants to turn green in the spring, long before the willows lining the riverbanks had tentatively ventured forth with their golden green leaves, and while clouds of pollen from still-leafless elms were being gently scattered by March winds.

Centuries ago, during one of its annual and uncounted March floods, the Platte overflowed its low banks. Its surging waters spread out aimlessly over the sodden prairie, carrying large ice floes that prodded and ripped at the prairie plants. Under these forces the grasses and sod were eventually torn asunder and carried downstream with the water. Thus, a new river channel began to form, isolating an area on the north side of the main channel of the Platte that extended some sixteen miles from the point where the river originally overflowed its banks to where it found its way back to its original channel. In this way a large elongated island was formed. Subsequently the island was to become dissected by stream-cutting action into an upper and a lower portion. The upstream section was eventually to become known as Shoemaker Island; the lower one became Mormon Island.

As the spring floodwaters slowly subsided, the prairie grasses growing along the new channel were suddenly faced with a harsh environment. Silts and sands had buried them as the receding river gave up its sediments to the land, and most of the prairie vegetation, unable to cope with such conditions, soon died. Mixed among the sediments left by the river were also the tiny seeds of willows and cottonwoods, which quickly germinated in the spring sun and the moist surroundings. Clinging tightly to their shaky substrate, the seedlings fought a silent race with time to stabilize the land before both they and it were swept away in the next spring's flood. Supported by annual weeds and grasses whose seeds had also washed ashore, the vegetation somehow held the land together that first year. The seedlings of the willows and cottonwoods were soon reinforced by those of elms and ash, whose somewhat larger seeds had been carried in on the wind by their winglike appendages or washed in by the river.

For the first four or five years the tree seedlings competed strongly with annual grasses and weeds. But each year more silts and sands were deposited, often burying the lower vegetation while permitting the taller woody seedlings of shrubs and trees to survive. Shoreline shrubs such as dogwoods began their growth a few years later than did the earliest trees because their larger and rarer seeds were only occasionally carried downstream by the river or brought in by birds. Of all the trees, the willows and cottonwoods grew fastest in the low-nutrient and nearly water-saturated soil environment. Yet, as the organic matter and nutrients of the soils increased, many other plant species also invaded the area.

Some fifteen years or so after the initial flooding, the new river bank was jammed with shrubby vegetation, and the sandbar willows were already nearly twenty feet tall. These slender trees formed a sturdy barrier between the river and the land; their intertwined roots held the soil firmly in place and their closely spaced stems and canopies shaded the soil below. Indeed, as the soil and debris built up around their roots, the drier conditions thus produced placed the willows and shoreline shrubs at a disadvantage. Finally, they began to die, only to be replaced by other plant species better adapted to surviving under these new conditions. Now the cottonwoods began their ascendancy. They soon outstripped the willows and other shoreline vegetation as they began to build an ever firmer bulwark against the river. By the time a half-century had passed, the cottonwoods dominated the shoreline scene. Yet even they began to mature and slowly die as the island approached a century of separate existence. Finally, the more slowly growing ashes, elms, and hackberries added their stature to the narrow forest, and a rich array of flowering herbs flourished at their bases.

As the island approached the end of its first century, it had acquired a distinctive appearance. Like the prow of a great sailing ship headed into the surf, it was crowned at its upstream end with giant cottonwoods and elms. For many decades they had borne the brunt of the yearly ice floes and stood well above summer water levels as they had gathered around their broad bases the sediments of a hundred springs. Lining the side channels of the island was a narrow but rich forest of ashes, boxelders, mulberries, and hackberries, as well as occasional cottonwoods and elms. Between them and the water's edge was a protective woody border of willows and shrubby dogwoods. Inside the wooded edges of the island was a large area of subirrigated native meadow, which was punctuated here and there by a tracery of trees or shrubs that still marked an old filled-in stream channel. Here too were irregular depressions that in spring filled with snow melt and which supported a profuse growth of cattails and other marsh plants. At the lower end of the island, shrubs trailed off into a sandy bar, where sandbar willow seedlings constantly engaged in their endless reclamation battles with the river.

By the time of the great immigrant influx, the darkly wooded island was a complete entity unto itself. The trails leading west from Nebraska City were well isolated from it by the broadest and southernmost of the Platte's major channels. Toward the north, it was separated from the Mormon Trail and Grand Island proper by the much narrower middle channel, which could be easily forded. Thus, the north side of the the island was first influenced by the immigrants. Initially it was used for little more than summer cattle grazing on the rich meadows of its interior, but as bridges began to be constructed across the many small subdivisions of the erratic middle channel, the island was increasingly opened for homesteading.

As the Union Pacific railroad approached Grand Island from the east in the mid-1860s, these bridges provided a convenient means for timber cutters to invade Shoemaker Island, Mormon Island, and other similar islands to cut the shoreline forest. The taller trees along the other banks of the Platte had long since either been chopped down for construction or firewood or perhaps had been destroyed by periodic prairie fires. So the trees of the islands were prime targets for the railroad contractors, who ruthlessly invaded the lands of squatters and homesteaders and cut down all the trees large enough to use for railroad ties.

The south channel of the Platte was finally bridged successfully below Grand Island in 1874, and thereafter travel was unimpeded through Mormon Island to the region south of the river, where the town of Doniphan was founded in 1879. The area between Grand Island and Doniphan became known as the "Nine bridges," since that many wooden bridges were needed to cross all of the large and small channels of the Platte in that region. With the construction of these and other bridges farther upstream, linking Shoemaker Island with the mainland, Mormon and Shoemaker Islands began to lose their unique identities, and their prairie meadows were broken for cultivation. The river continued to rage every spring, flooding the lower parts of the islands and sometimes sweeping away livestock with it. Yet, each spring also brought the flocks of sandhill and whooping cranes that roosted nightly on sandy bars and islets annually swept clean of vegetation by the floodwaters, and the channels of the Platte were alive with the hordes of migrating geese and ducks that waited for the rivers and lakes farther north to become ice-free.

An immature bald eagle coursing above the river, over flocks of wintering mallards and other waterfowl. Photo by author

Now I climbed the hill and arrived at the top,
There where the streams of water shorten people's legs,
Where once great herds of buffalo started from,
Yet I saw nothing yonder but an expanse of land.
Only the expanse of land, only the expanse of land.

Pawnee song (Densmore, 1929)

The Present

Piping plover nest on a Platte sandbar. Photo by William S. Whitney, Flatland Impressions, Aurora, Nebraska

The Sandbar

The low island lay barren and lifeless in the cold February sunlight, which was slowly melting the last remnants of snow that had been deposited a few days before and a few scattered chunks of ice that had been rudely thrust upon it as the upper Platte had broken up. Indeed a few pieces still occasionally floated past the bar after having been hung up for a time in shallows or caught by shoreline debris. The sky was finally clearing, the wind was shifting to the south, and the lowering sun was occasionally casting long cold rays of light through the broken clouds and across the western sky.

Faintly at first, but then with ever increasing clarity, a babbling sound became apparent to the south, and a thin, wavering line of birds materialized on the southern horizon. As they approached the river the birds did not perform their usual wary circling. Instead, almost without hesitation, they set their wings, lowered their long legs, and pitched downward to the waiting bar. They were lesser sandhill cranes, and they had started north the day before from northern Texas and flown nonstop since leaving Muleshoe National Wildlife Refuge some 550 miles to the south. After taking long drinks and shuffling and preening their feathers for some time, the birds waded out into the shallow water at the upstream end of the bar, and gradually quieted down to rest.

The cranes were the first vanguards of a group of nearly 250,000 birds that would soon be arriving in the Platte Valley, as they had done each spring for uncounted millennia. By virtue of its wide and shallow channels, abundant sandy islands and bars, and scarcity of heavy vegetation on the more recently formed bars, the river offered perfect nighttime protection from coyotes, dogs, and other land-based predators. The nearby lands provided a mixture of native wet meadows rich in earthworms and insects and an abundance of corn-stubble fields that still carried a rich trove of loose, scattered kernels and occasional unharvested ears. For the next month the cranes would alternate flights to the river each evening for roosting, and to the fields and meadows each dawn for foraging. In the process, they would gain up to a half-ounce of fat per day, or nearly an additional pound of weight during the month-long period in the Platte Valley. Another pound or two would be acquired in the northern part of the Great Plains on later stopovers before the cranes left Saskatchewan for the rigors of the arctic spring and the later stresses associated with breeding.

The bar was ideal for the cranes. It was well isolated by deeper water on both of its elongated sides, and the bar itself had large areas of barren sand and silt that had recently been deposited by the temporary flood levels of the river following ice breakup. As the early days of March passed, more cranes arrived, and so too did vast flocks of arctic-breeding Canada geese and white-fronted geese, many of which alternated their evening roosting activities between the river and many large "lagoons" in the clay bottom lands to the south. As these lagoons were drained or when low winter snowfall kept their levels low and less suitable for the geese, the geese spent more and more time on the river, usually roosting well away from the crane flocks, which preempted the better sandbars for their own use. At times the cranes and geese were temporarily associated, as they were when eagles flushed them from adjacent roosting, sending the flocks into confusion and frantic clouds of swirling birds. Paired birds and their youngsters often became separated from flock or family and would join goose flocks for a time; sometimes the reverse happened.

By the end of March all of the geese had passed to the north and about a week later, in early April, a morning breeze from the south and warming temperatures set the cranes into motion. Flock after flock arose from the river, circling ever higher in great circles, until they were several thousand feet above land. Then, as if on cue, they struck out directly north, leaving only footprints and a few feathers behind on the island as temporary proof of their stopover there.

As the cranes left, their places were taken by migratory shore birds such as semipalmated plovers; common snipes; pectoral, Baird's and semipalmated sandpipers; and greater and lesser yellowlegs, all of which stopped for a few days while on their way north. But killdeers and spotted sandpipers soon set up territories on river shorelines and islands, and in late April several piping plovers appeared as well. One of the pairs became territorial on the sandbar, while another occupied a similar site a few hundred yards downstream.

Now too along the river Franklin's gulls and Forster's terns appeared, the gulls gathering in nearby fields as farmers began to till their lands, and the terns gracefully cutting the air along the river's channels, seeking out killifish and other surface-feeding minnows. By mid-May they were supplemented by black terns, which stopped only for a few days to feed on emerging insects before passing on to breeding rounds in the Sandhills marshes to the north. The last of the terns to arrive were the least terns, six of which appeared the last week in May and quickly settled in for the summer.

Only a few days after they arrived, the least terns began courting. About the island and along the channels beside the land a tern might fly with a minnow it had just caught; one or more birds would give chase and all would scream excitedly. At the end of the flight the bird would begin a silent glide, with its wings held up in a distinct V, and

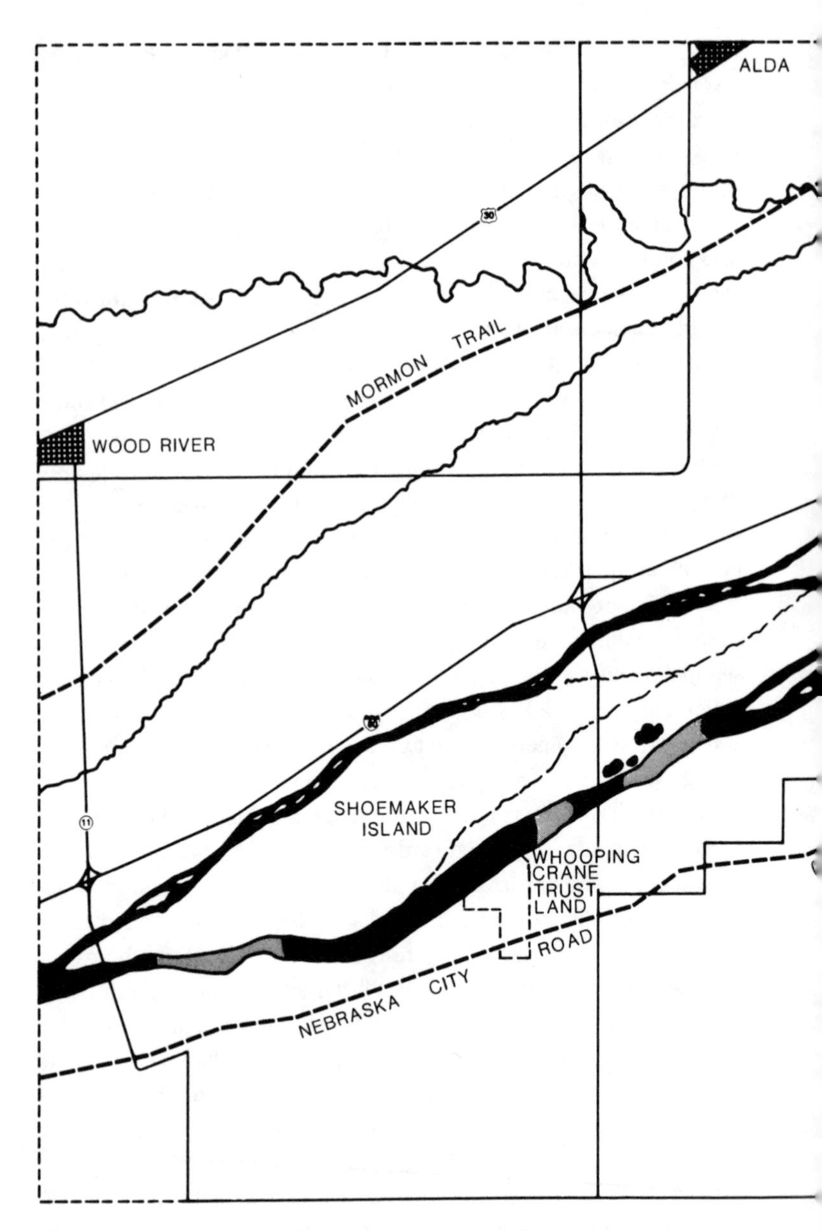

FIGURE 2: Map of the Platte River near Grand Island, Nebraska, showing Shoemaker and Mormon islands and areas of the river of particular importance to sandhill cranes.

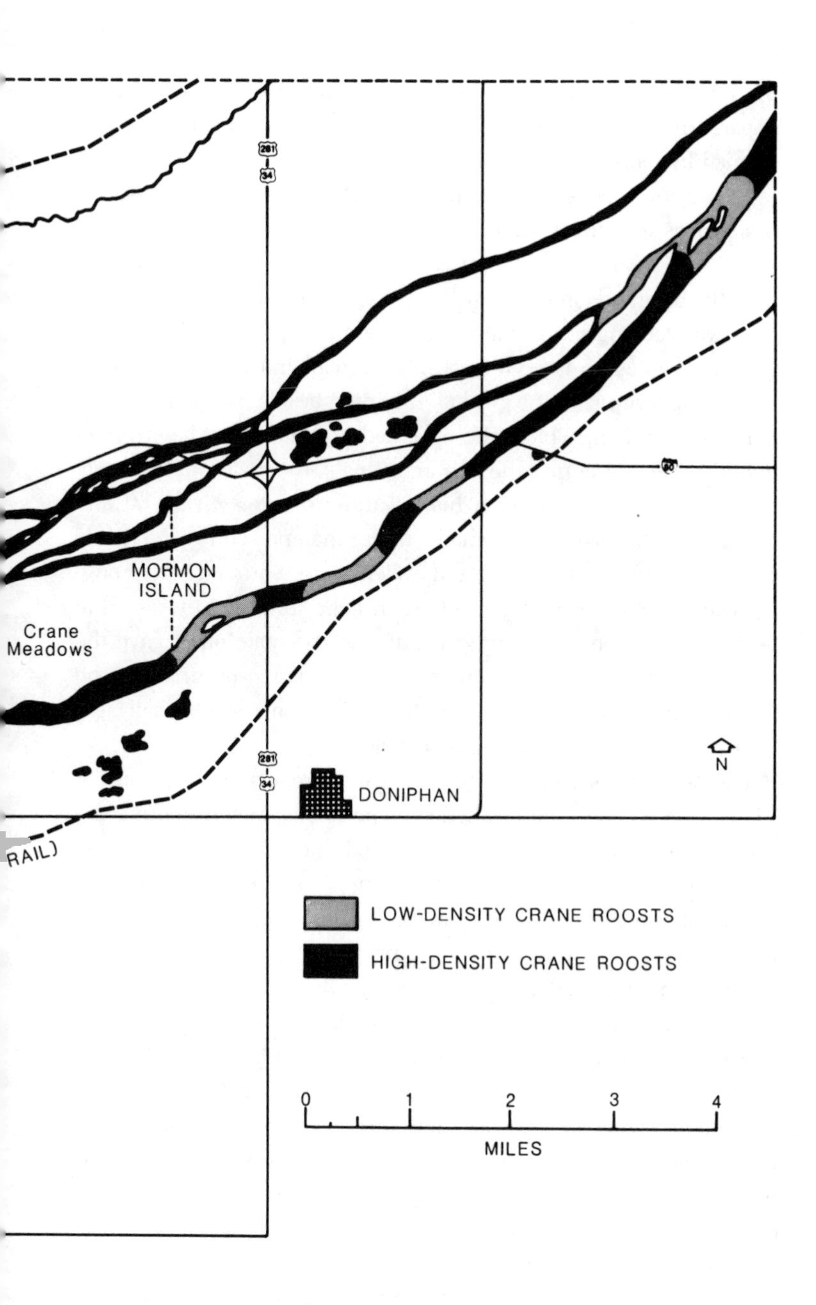

MORMON
ISLAND
Crane
Meadows
DONIPHAN
N
RAIL)
LOW-DENSITY CRANE ROOSTS
HIGH-DENSITY CRANE ROOSTS
0
1
2
3
4
MILES

the one behind it following in the same way. During these flights apparently pair bonds were initially formed, and pair recognition was learned by individual differences in calls.

After a few weeks of excited fish flights, the six birds had sorted themselves into three separate pairs; these spread out on the long island, one near the upstream end, one at the middle, and the last near the downstream tip. Now the second phase of courtship began, courtship feeding of the female. As the male approached in the air with a minnow, he uttered his fish-flight call and was answered by the female waiting on the ground. Alighting beside her, he presented her with the fish, which she accepted and ate immediately. Frequently after presenting the fish to his mate, the male would elevate his wings diagonally and tilt his bill upward before flying off and seeking another fish. Occasionally, rather than presenting the fish to the female immediately, he would walk in a circle around her, strongly raising and lowering his head. Such behavior evidently served to initiate copulation, and during copulation the male often gave the female a fish. However, at other times copulation occurred in the absence of an actual fish, with the birds simply symbolically exchanging fish by interlocking their bills.

The nest sites selected by the terns were vegetation-free areas of sand, simply consisting of shallow depressions. Where natural depressions were available, these were used; but otherwise the female simply scratched out a shallow cuplike area by kicking with her feet and worked her body into it to make it fit more perfectly. Near one least tern's nest was that of a piping plover, which likewise was nothing more than a depression made by a deer's hoofprint. The pale color of the bird's back blended so well with its sandy surroundings that it simply appeared to be a part of the island itself.

The eggs of the small colony all were laid during the end of May and the first week of June. By then, the river's level had fallen even farther, leaving the nests well away from the nearest water and even more safe from possible flooding. Eggs were deposited daily, until two or three eggs were present in all of the nests. As soon as the first were laid the female began incubation, with only occasional participation by the male. Thus, most of the time the male spent in fishing and bringing food to his mate, which he fed many times a day.

During the middle of June, as the daytime temperatures soared into the high nineties, the incubating bird spent much of its time simply shading the eggs, and sometimes it even flew off a short distance to soak its breast and belly feathers in water. Returning to the nest, it would settle down briefly to wet the eggs, then resume its stance.

Hatching began three weeks after the eggs had been laid, during the third week of June. The chicks hatched on succeeding days, and within a few hours after hatching they were fed by the parents. Soon after hatching was completed the adults also picked up the eggshells, flew off with them about twenty-five yards, and dropped them in the river. The males were kept continuously busy from the time the nests hatched. They captured small killifish, returned to the nest, and fed the young directly or passed the food to the female, who either fed it to the chicks or ate it herself. After only a few days the downy chicks began to wander away from the nest site, often walking toward a clump of weeds where they would sit quietly in the shade until it was time to be fed again. They grew rapidly, and in less than three weeks they were beginning to practice trial flights. Yet they were still dependent on their parents for food, and although they occasionally tried their luck at fishing, they were so awkward that they rarely were successful.

By then too the piping plover nest had hatched, and two of the four chicks had successfully survived long enough to fledge. Other than the piping plover and the least terns, no land animals bred on the barren sandbar that summer except for some six-lined racerunners, which had hibernated several inches deep in the sand over winter. They had not begun to emerge until early May, when the daytime temperatures became warm enough for the animals to move about rapidly in search of grasshoppers, crickets, beetles, and other insects that they found by their keen sense of smell as much by their eyesight. Both the sexes of the adult lizards were beautifully marked with six greenish yellow racing stripes extending down their backs, but in adult males these stripes had gradually almost disappeared with age, while the back and sides had become progressively suffused with greenish coloration, shading into a rich bluish green laterally.

The female racerunner's quarter-acre home range overlapped that of the least tern colony, but she was not bothered by them, although they often noticed her basking in the sun on the warm late-spring mornings. She would typically leave her burrow in mid-morning and spend an hour or two hunting by probing under leaves and into small holes or pawing at surface vegetation, occasionally sticking out her long forked tongue as if tasting the air for scent of prey. Whenever she caught a large grasshopper nymph she would kill it by crushing its thorax, shake off its legs, and then consume it greedily. Later, if hunting was not favorable, she might return and eat the legs. On one such morning in late May the female was suddenly surprised by the sight of an adult male, whose flashing blue sides attracted her eye. He had a broader head but a slimmer body than she, and he was easily recognized as a male by his behavior. On seeing her, he rubbed his cloacal area from side to side on the sand, and vibrated his long tail rapidly. He then quickly approached the female, who seemed mesmerized by him, and quickly grasped her on the top of the neck. In this position the female carried the male about for a short time, as the male arched his body and tried to bring his vent into contact with hers. Mating was finally accomplished, after which the female broke free of the male's grasp and darted away, while he remained basking in the sun for a time.

A few days later the female began digging a nest, which simply consisted of a tunnel in the sand, with domed ceiling and an entrance that was closed off by the female after she laid her eggs. The size of the cavity was substantially larger than necessary to accommodate the three large eggs that she laid in the moist sand, nearly a foot below the surface. As the eggs developed they gradually absorbed moisture and more than doubled in weight, although in length they increased but slightly. After laying her clutch and closing the nest site off, the female never returned to it, leaving its fate to chance. The eggs developed without harm, however, and in early August the new hatchlings emerged into the sunshine. They were only a bit more than an inch long at hatching and weighed only about a gram or less, or about a tenth of the weight of their parents. Yet, those that survived their early days grew rapidly, and by the time of their fall hibernation some six weeks later they were nearly half as long and as heavy as adults. During that time only their quick speed saved many of them from being caught by coyotes, raccoons, skunks, and perhaps even grasshopper mice.

The Shoreline Forest

The century-old, archaic-looking cottonwood that guarded the upstream tip of the island was totally bare of leaves and stood gaunt against the gray sky. It was surrounded by a retinue of lesser trees that formed a protective barrier around the island, and for most of the winter it had been a regular daytime perching place for an adult bald eagle. She was a large female that had nested that previous summer in northern Minnesota and had gradually drifted south as winter had progressively shut down fishing opportunities farther north.

The middle Platte had also largely frozen over, but a small area upstream in the vicinity of Lexington had remained ice-free where the outlet of the Tri-County Canal churned and slightly warmed the water. There the female and several dozen other eagles had spent the coldest months, feasting on a variety of common rough fishes such as carp, redhorse, catfish, gizzard shad, and carpsuckers, as well as the carrion of many rodents, rabbits, domesticated livestock, and waterfowl that had died of hunting wounds or other mishaps. Of all of these, the gizzard shad had been the eagle's favorite food whenever the water temperature had risen sufficiently high above freezing to allow these fish to feed near the surface, or when they had become stunned or killed by passing through the power turbines of the Tri-County Canal. Like the waterfowl, the eagles gathered wherever there was open water, and now in early March the entire river was essentially ice-free. As a result, both the eagles and the waterfowl had scattered out over the entire Platte valley between Lake McConaughy and Grand Island.

As the eagle gazed up and downstream from its commanding perch, it noted without apparent interest the flights of courting pintails and mallards. It soon let its eyes come to rest on a roosting flock of small Canada geese that had gathered a half-mile upstream after having arrived late the evening before. The eagle had not yet made its morning patrol, but now, as the day was warming and updrafts were developing, it lazily slipped off its perch and headed directly up the river, sending knots of green-winged teal and pintails into panicky flight before it. The geese sent out a warning call and also frantically beat their way into the air, forming a confused mass that the eagle quickly assessed. One of the birds was clearly having trouble gaining altitude. Further, it was dragging one leg as a result of a buckshot wound it had suffered the month before in Texas, and it was only now beginning to recover. Ignoring all the other geese, the eagle plummeted downward through the flock, striking the crippled bird with all of its talons and throwing it into a head-over-heels tailspin down toward the river below. Checking its flight, the eagle wheeled and quickly coasted down to the dying bird. The goose weighed more than four pounds, or more than the eagle could easily lift, so it began tearing the carcass to pieces right on the edge of the island where it had fallen. Another eagle, flying slowly upstream, saw the activity and approached to join it. However, it was a much smaller juvenile male bird, and it was easily kept at a distance by the strong and dominant female. Finally, having eaten its fill, the female took off and slowly flew back to the cottonwood, leaving the remains of the goose to the waiting young eagle and a few crows that had also gathered to watch.

Not far from the tall cottonwood stood a tall elm, which thus far at least had survived the Dutch elm disease without apparent harm. In a large crotch a pair of crows had nested two years before, but both had eventually died after contracting avian cholera from feeding on diseased birds in the Rainwater Basin to the south. Thus, the nest had remained abandoned until it had been taken over during the winter by a pair of great horned owls. The owls were permanent residents of the island, and they lived off a wide variety of prey, including many small rodents, rabbits, and birds up to the size of mallards, especially if the latter were in some way incapacitated. Early in February the pair had set about a limited renovation of the nest, which stood some thirty feet above the ground and was well supported from below by heavy branches. Now, in early March, the first of the three white eggs was hatching, nearly a month after it had been laid. Still to hatch were the remaining two, which had been laid at approximately three-day intervals after the first. Most of the incubation had been done by the larger female, while her mate hunted for both of them. Leaving the woods each evening at sundown, he would course the river, initially looking for vulnerable birds, then turning his attention to nocturnal rodents as they began to make their appearance. Several times each night he would noiselessly return with prey to feed to his waiting mate, and as daylight approached he returned to his waiting post some distance away from the nest. There he would inconspicuously sit through the daylight hours, trying to avoid being seen by crows or bluejays, which harrassed him incessantly whenever he was discovered by them. In a few cases he had gained revenge on the crows by ghosting into their winter roosts and slaughtering one of their number, but now most of the crows had left the valley and were heading north, leaving this island largely to the owls.

The tiny owlet in the nest was a fluffy ball of yellowish white down; even now at two days old its eyes were still closed, and it was barely able to stand up. Yet it would have a distinct advantage over its nest mates, for by the time they had both hatched it would have at least six days' growth behind it, which would place it in good stead for demanding its share of the food that the adults would bring. In most years this was not a real problem for the pair, for the Platte had an abundance of prey items that nearly always allowed enough for all the owlets. Yet at times a heavy snow cover and frozen river reduced both the bird population and the ease of finding small rodents, which tended to remain below the snow, making hunting especially difficult.

Not far downstream from the owl nest was a large elm that had died nearly ten years before, when Dutch elm disease invaded eastern Nebraska and marched westward along all the major river valleys. It had decimated the shade tree populations in cities and stimulated an enormous outcry of concern by the populace. The cities responded to pesticide interests by blanketing the towns with clouds of DDT, killing songbirds by the thousands and poisoning the fishes in nearby streams, but doing nothing to prevent the spread of the disease. Only several years later, when the towns were nearly devoid of elm trees and DDT levels were reaching alarming levels in human and animal tissues, and the federal government finally and reluctantly decided that it must act to restrict the sale of the insecticide within the United States (but not its export to other countries), did the madness subside. Slowly the robin and other town-adapted songbirds recovered. Their populations in cities were also gradually supplemented by rural populations, which luckily had been spared the intense chemical warfare that occurred in the cities and towns.

A woodpecker excavation begun several decades before had been used for years by flickers, and later had been enlarged and taken over by fox squirrels. With the dying of the tree, the squirrels abandoned it and moved to a nearby one that offered better summertime shade and cover. The most recent occupant of the decaying tree was a female wood duck. Wood ducks were a fairly recent arrival in the central Platte. A century ago there were virtually no trees large enough on the Platte to provide adequate-sized nest cavities for them, and they were confined largely to the Missouri flood plain. Yet, with the planting of trees and the growing up of a mature riverine forest along the Platte the birds slowly moved west, and by the late 1900s they had nearly crossed the entire state as breeding birds.

The pair of wood ducks had returned to the area in mid-March, slightly later than the mallards and pintails, and immediately moved onto some treeshaded pools that marked the site of an extinct river channel. The pair had already finished courting activities on their wintering areas in southern Missouri, but even now the male remained highly attentive to his mate, frequently flashing his beautiful iridescent wing speculum pattern to her in a mock preening display, swimming proudly ahead of her and orienting his long, flowing nape toward her, or directly facing her and extending his long neck, exposing his white-spangled maroon breast and immaculate white throat to her view.

By late March the pair had carefully examined all the large trees in the area, flying from one to another and clambering about the branches in search of suitable nesting cavities. The male took as much part in this search as did the female, and on finding a possible nest site he would utter a soft call and direct the female toward it with bill movements. She would then approach it, enter it to examine its interior, and then pass on to the next possible site. This "house-hunting" activity occupied several days, for the birds needed a location that was fairly high and spacious inside, but with an entrance small enough to keep raccoons out. A lateral opening, rather than a vertical one that tends to let rain in, was also important, and ideally the tree should either be situated immediately above water or very close to it. Trees standing in water would tend to keep terrestrial predators away and furthermore would not require long walks to water by the newly hatched ducklings after they emerged from the nest.

The cavity that was finally selected by the wood duck female was about four inches in diameter and about two feet deep internally. It was only about ten inches wide inside, or so narrow that the female's long tail had to be cocked upward as she sat on its floor. The female gradually lined the bottom of the nest with a mixture of down and breast feathers while she was also laying a clutch of eggs, at the rate of an egg per day. When not actually laying eggs, which she did early each morning, she and her mate spent the day foraging. On their wintering areas they had feasted on acorns, their favorite food, but here there were no oaks to be found, nor were there hickory nuts or beechnuts. Thus the birds looked for seeds of aquatic plants and ate the green parts of aquatic plants such as duckweeds and aquatic buttercups that were growing in the small pool near their nesting tree.

Once the female had completed her clutch and begun incubation, the role of the male in reproduction was over, and he eventually moved out of the nesting area. He found a number of other post-reproductive males, and they moved downstream as a group. They soon reached the Missouri River and thus were assured a future supply of acorns and hickory nuts that would ripen during the summer months.

For her part, the female became extremely secretive. She left the nest only a few times a day, just long enough to drink and feed for a short time before returning to the nest. She usually undertook this flight in early morning and late afternoon, and for the first two weeks of incubation the male had accompanied her. After his departure she made these flights alone and silently. On her return she would weave through the forest vegetation like a guided missile, darting and turning to avoid branches, and flying full speed to her nesting tree. There she very suddenly would brake with wings and tail and would quickly disappear into the nest cavity. There too she was completely safe from the great horned owl and nearly all other predators, and the month-long incubation period proceeded without interruption.

In mid-May, when the forest was well shaded with new vegetation, and insect life was present everywhere, the clutch of thirteen eggs hatched. Incubation had begun at the completion of her clutch, and thus all the eggs began pipping at about the same time. Over the next six hours all of the eggs hatched, leaving little room in the nest for both the female and thirteen squirming youngsters. But by then it was late afternoon, and the female brooded the chicks tightly until the next morning. Then, at about sunrise, she flew down out of her nest and landed at its base. Uttering an insistent series of low *kuk* notes, one chick after another clambered up the inside of their nesting cavity, appeared at the hole, and without the slightest hesitation launched themselves into the air. One or two bounded against branches as they half fell, half flew downwards, beating their stubby wing-buds frantically. Yet, they all landed without mishap, and after a few minutes of rounding them all up, the female headed for the nearest pool, her long string of ducklings following closely behind.

In the next few days her brood gradually diminished. A few became separated from their mother and died from exposure or lack of food, two were eaten by snapping turtles, and one was consumed by a bass that had become isolated in a pool and was itself destined to die that summer as the pool gradually dried up. By the time the brood was approaching two months of age their primary feathers were nearly grown, and their numbers had been reduced to five. When they were eight weeks old and about ready to fledge the female had deserted them and had begun her own summer molt. In a few days her primary and secondary feathers fell out, and she became totally flightless for a month. At that time she remained in heavy cover, scarcely moving about at all, until her newly growing flight feathers offered her the capability of again escaping from danger by flight.

Summer thus passed, and the birds of the shoreline forest fledged their young, in some cases only to begin again on a second brood. Such were the mourning doves, which persistently bred time and again for as long as the summer period lasted, producing a total of four successful broods and a few failures before finally ending their efforts in early September.

A wet meadow of tall prairie grasses, bounded by a leafless elm and smaller trees in late March. Photo by author

The Wet Meadow

When the male red-winged blackbird returned to his marshy territory of the previous year, he found the water still frozen and the tall cattail heads on which he had perched so many times beaten down by the winter's snows. For lack of a better place, he adopted a fence post as his prime territorial perch and began his first, halfhearted attempts at territorial song and advertising display of the season. In a sense, he needn't have bothered. Few other male redwings had yet returned, and it would be a week or more before the females would begin appearing on the small piece of marshy grassland that cut through the island and represented an extinct flood channel of the Platte.

Now the river only rarely rose far enough to actually cause a surface flow through the channel, but instead a subterranean water table brought moisture up from below. This produced a stillwater habitat that grew up to tall prairie grasses such as Canada wild rye, cordgrass, and big bluestem in late summer, and also a few cattails where the water stood well into the summer months.

The hardy redwing was the first of the territorial occupants of the area, but it had already been used for some weeks by the sandhill cranes. Although the cranes now spent the majority of their foraging hours in cornfields, the wet meadows were their traditional feeding grounds in the Platte Valley. These lands offered not only close proximity to the river, but also provided at times a type of secondary roosting site, where the birds gathered at dusk before flying to the river and, at dawn, before heading out toward the more distant grain fields. The low meadows also offered such protein-rich foods as earthworms, grasshoppers, spiders, and adult or larval beetles, as well as a small amount of green leafy material from newly germinating plant life.

The male redwing, being one of the very first of its species to return to the island, had its pick of the entire area for choosing a territory. The site that it selected bordered a gravel road, where a fairly deep ditch normally held water well into the summer and supported a thick border of cattails, phramites, and cordgrass. Other aquatic vegetation graded out into the wet meadow behind the site; this area held a thick growth of sedges mixed with wild rye and other low prairie species. Here the water supply was less certain, since it came as much from underground seepage as from spring runoff. But the land was much too wet to plow, and thus it served well for cattle pasturing throughout the summer months.

Of the various tall plants available to the redwings, the cattails were perhaps the most important. Not only did they make excellent song perches but they were also sturdy enough to support the large nests that were later to be built by the females. For nearly a week the male redwings sang and displayed their red epaulets toward one another by "song-spread" displays or by short flights above their territories, while keeping the blood-red covert feathers raised. The song uttered during flight was rather different from the *gurga-lee* that was produced while perching. It was a long and rapid series of repeated *teee* or *chee* notes, usually uttered after the male had successfully repelled another male from his territory. Simultaneously, the epaulets would be fully raised, the tail spread and lowered, and the wings beaten so slowly that the bird would sometimes almost seem ready to stall in flight. Occasionally a rather different "fluttering flight" would also be performed, especially early in the season as the first females were arriving. In this version the tail would not be spread and lowered, and the wingbeats were much more shallow and rapid.

But much the most common display was the song-spread, during which the male would suddenly fall forward on its perch, raise its epaulet feathers, spread its tail, fluff its head, nape, and back feathers, and arch its wing outward and downward. This display was often directed toward another rival male, but at other times it was directed toward a potential female mate. In this latter situation the display was always intense and would take on a special form, in which the head is hunched, the tail is lowered and spread, and the wings are drooped and held in this extreme position for extended periods of time. In this crouched posture the bird would sometimes utter a repeated note and begin to flutter his wings; as he approached the watching female he would flutter his wings conspicuously. In this context the display became an invitation to copulation and, if receptive, the female would hold her body horizontally, flex her legs, and elevate her tail while uttering similar soft notes. Mating would follow immediately, and sometimes several mating attempts would occur in rather rapid succession.

As the females arrived and became mated, the males would frequently concentrate their displays at potential nest sites, such as a cattail clump. After a short flight-song the bird would land on the cattails, raise his wings high above his back, and point downward with his bill into the clump of vegetation. Sometimes females would follow the males into that clump of vegetation, seemingly evaluating it themselves as a possible nest site. Yet, the final choice was hers, and although males would sometimes engage in "symbolic" nest building by picking up bits of vegetation, all of the actual nest construction would be done by the female in a site selected by her.

The pair bond of the birds was a rather informal and tenuous one at best, beginning when the female entered the male's territory and apparently accepted both him and it. From that time until the completion of a nest and the start of egg laying only about three weeks would elapse, and no sooner would a male have begun a nest than he would turn his attention to a new female. Some males would thus accumulate a harem of several females on their small territories, with the number of females attracted to a particular male probably being related to the richness of the territory in terms of potential nesting sites rather than size. Since the females tended to select cattail stands occurring on the edges of open water, the ditchside location of the territory picked by the first male to return was perhaps responsible for the fact that no less than three females were fertilized by him. Each of these was slightly out of phase with the others in its breeding cycle, for, like the dickcissels nesting in the nearby prairie, the male was essentially forced to court them one at a time. Also in common with the dickcissels, the nests of the redwings were easily found by female cowbirds, and virtually all of the nests suffered the fate of having at least one cowbird egg deposited within them. In this case, the eggs of the redwings were distinctly larger and heavily striped, rather than spotted, with brown, but nevertheless they were regularly accepted by the redwing females. However, the young cowbirds were scarcely separable from the baby redwings after they hatched, including their gape and bill-flange coloration. So they were well tended by the female redwings, who thus often neglected their own youngsters.

Scattered here and there among the cattails were a few very different-looking nests attached to cattail stalks. They looked like miniature footballs and were about seven inches high and five inches wide, with a small rounded opening just slightly above the midpoint. They were made of strips of cattail leaves woven together and to the living cattail stalks in such a way as to adhere tightly to the vegetational substrate on which they were placed. In a small area encompassing only about fifty by one hundred feet no less than six such nests were present, all built by a single male. The industrious bird responsible for all this construction was a male long-billed marsh wren, which was still hard at work constructing his seventh such nest. Each required only about three days to construct, and the entire group of six nests had thus taken less than three weeks to build.

In spite of this tremendous housing supply, the male had thus far acquired only a single mate. Indeed, his nests had several functions other than serving as egg repositories, and one of these was to mark his courting center. He would sing from this small cluster of nests during most of the day when he was not either foraging or nest-building, and he would sleep in one at night. Periodically a female would be attracted to the area and investigate one nest after another. The male would show them to her in sequence, much in the manner of a realtor taking a client on a house-inspection tour. Once the female had selected the male as a mate, she would either accept one of the nests as a breeding nest—after a partial redecoration that involved the insertion of an inner nest lining of feathers or strips of cattail stalks—or she might reject all of the nests completely and begin building one of her own from scratch, which the male then would take over and complete himself, except for the final lining.

As soon as the male had acquired his first mate and she was occupied with her own egg-laying duties, he shifted to another area of his territory and began a new round of of nest building in a different courtship center. Eventually he would construct nearly twenty nests and acquire a total of three mates that nested within his territory. Unlike the nests of the blackbirds, those of the wren were too small to allow female cowbirds to enter, and furthermore their enclosed feature kept out rain as well as protecting the clutches of eggs during severe windstorms.

By the time the male had acquired his third mate, the young of the first nest were hatched, and from that point on, the male had no time for further courtship. Instead, he was forced to gather food from dawn to dusk, feeding the young of one nest after another. Females also fed the young of their own nest, and both sexes would remove the fecal wastes of the young birds until they were about eleven or twelve days old. After that, the young simply eliminated their wastes out over the edge of the nest. A few days later they were able to leave the nest on their own, but the parents continued to feed them at reduced rates for another two weeks. By that time the juveniles of several broods began to amalgamate, and the youngsters moved around the marshy area on their own. With their independence, the male again began a new round of nest building and courting activities. By the time the summer ended, several dozen nests were scattered throughout the wet meadow. Some of these would be taken over by mice, others by bumble bees, and yet others would be used for late fall or early spring shelters by adult wrens well beyond the time they served any reproductive role.

As the young marsh wrens wandered about the wet meadow, they became easy prey for various predators, including a marsh hawk that regularly coursed low over the meadow in search of rodents or any other available food that it might surprise during its low foraging flights just above grass level. It too was hosting more than a single mate; two females nested fairly close to one another within his territory, and the male was kept hunting virtually all day long in an effort to feed both of his mates and their young broods. Although mice constituted the largest single component of his prey, he did not hesitate to take frogs, garter snakes, songbirds that he could catch, or even occasional small ducks such as blue-winged teal that he sometimes surprised on their nests.

Yet a substantial number of young redwings and marsh wrens survived the summer, together with several juvenile bobolinks and yellow warblers that had also been hatched and reared in the meadow. By the end of August, as the meadow dried out, the birds that had been breeding on it gradually moved away from the area. The redwings began to merge with others of their species, as well as starlings, cowbirds, and grackles, and began to move out into the ripening grain fields. By late summer the meadow was totally dry and relatively devoid of birds, save for the daily cruises of the marsh hawks, which were still engaged in feeding their nearly grown broods.

The Wood Lot

A short distance from the river, near a deserted farmhouse and barn, was a rectangular wood lot that had been planted in the thirties as a shelterbelt and a windbreak for the farmyard. It was now fully grown, and indeed some of the trees were beginning to die, but yet it provided a dense, woodslike environment that was rich in animal life. A bobwhite quail covey roosted nightly along its brush edges, and more than two dozen bird species nested within its confines, producing the highest density of breeding birds to be found anywhere on the island. Mourning dove calls softened the spring air with their restful sounds, grackles nested colonially in the well-grown cedars along its edges, and the old nests of several pairs of northern orioles hung like forgotten Christmas ornaments from the tips of the lateral branches of the taller trees. The wood lot was also the undisputed domain of a pair of red-tailed hawks.

The hawks had used their nest in the tall hackberry tree for several years, adding to it a bit each year, and had recently relined their platformlike top with a few sprigs of cedars to ready them for the eggs. In mid-March, the female began her clutch, laying a total of three bluish white eggs that were heavily spotted with brown. Like the great horned owls nesting a few hundred yards away, she had begun incubation immediately, and by the end of April the first egg was hatching. Before egg laying, the two birds had patrolled their territories on a daily basis, chasing off all of the many redtails that were migrating through the Platte Valley during early March and sometimes engaging in aerial courtship. Soaring up several hundred feet, the two cut great circles in the sky, crossing and recrossing one another's paths as they ascended. Each time the male approached his mate he would lower his legs and adjust his flight so as to nearly touch her back. At these times the female would quickly flip over on her back at the critical moment, and the two would momentarily lock their talons. Then, both of them would partially close their wings and dive quickly downward, checking their flight only as they came close to their wood lot and swooping up again to repeat their performance. Often mating would follow such a flight, and it usually occurred on one of the tallest branches of a cottonwood tree growing along the river near the wood lot.

These spectacular aerial displays continued well on into the incubation period, when the male sometimes passed on a mouse or rat to the female in flight, but more typically he would bring it directly to the nest and feed his mate directly. Most of the prey were meadow voles, pocket mice, and cottontail rabbits, but at times he also brought in short-tailed shrews, fox squirrels, or garter snakes, depending on what happened to be available. Less often a dead bird would be brought back, such as a robin or a young chicken, and once the hawk even made an attempt to kill a feral tomcat that had itself been hunting for birds in the pasture. The stalking cat saw the shadow of the plummeting hawk only seconds before the bird's talons raked its back. Leaping into the air, the cat, with its quick reflexes, avoided a killing blow, and it struck back with its own razorlike claws. Suddenly finding itself in a situation it had not expected, the hawk released its grasp and struggled to escape the flailing claws of the cat. Both escaped serious damage, but the cat thereafter avoided the vicinity of the wood lot, and the hawk ignored the cat whenever their two hunting trails happened to cross.

Of all the birds breeding in the wood lot, few were more conspicuously abundant than the flickers and northern orioles. This is all the more remarkable since, like many other of the wood lot birds, they were relative newcomers to the central Platte Valley. Another feature that the flickers and orioles had in common and shared with a few other species was that each was actually a part of a species-pair complex, composed of eastern and western representatives. Probably as long ago as early Pleistocene times the ancestral flickers and orioles were separated and subsequently largely confined to eastern and western areas of North America, where their wooded habitats remained fairly unaffected by glaciation patterns. Later, as the glaciers retreated and the warm and dry postglacial period occurred, the arid plains provided few or no opportunities for range expansion. During this protracted period of relative isolation, which may have begun as early as the Pliocene, both the orioles and the flickers underwent local evolution and developed some distinctive characteristics. For example, the western population of flickers evolved a "red-shafted" appearance, owing to red carotinoid pigmentation being deposited in the wing and tail feathers of both sexes, and the males even acquired a reddish "mustache." In the eastern birds the wing and tail feathers retained or developed a golden yellow cast, and the males had black "mustaches." Likewise, males of the western population of Bullock's orioles became more yellowish orange throughout, including orange cheeks, and developed large white upper wing markings. In the eastern Baltimore orioles the more reddish orange carotinoid pigments prevailed in males; the entire head became or remained black throughout, and black pigment on the upper wings was more extensive, thus restricting the white areas to a small part of the upper wing surface.

With cooler and moister climates that followed the "climatic optimum" twenty-five hundred to seven thousand years ago, eastern species of forest trees penetrated the central plains by moving up river drainages. As these woodlands were supplemented by human-made shelterbelts and wood lots, the great barriers to forest-adapted birds on both sides of the plains began to crumble. Eastern forest or edge-adapted forms such as indigo buntings and rose-breasted grosbeaks moved westward and encountered similarly adapted western relatives in the form of lazuli buntings and black-headed grosbeaks. In these forms, as in the orioles and flickers, interactions between them resulted in competition and occasionally mixed pairing and hybridization. The probabilities of hybridization were of course related to the degree of genetic similarities and thus were greater in some pairs than in others. For example, in the flickers there appeared to be virtually no barriers to hybridization, while in the orioles there was somewhat less hybridization, and likewise the buntings and grosbeaks also responded with more limited mixed pairing.

In any event, the Platte Valley became one of the hotbeds of genetic contact between eastern and western birdlife, for it was one of the few river systems on the plains to span effectively the broad gap between the coniferous forests of the front ranges of the Rocky Mountains and the eastern deciduous forest riverbottoms of the Missouri Valley. Thus it was that the flickers of the wood lot and surrounding riverside forests were a surprisingly diverse group of birds. Many appeared to be to the eastern or yellow-shafted type. Others seemed to be of the salmon-tinted western type, and yet others were distinctly intermediate, the males sometimes with mustaches variably mottled with black and red.

Unlike the flickers, most of the northern orioles of the wood lot were of the eastern plumage type. The major area of east-west contact and hybridization was about two hundred miles farther west, in extreme western Nebraska and eastern Colorado, and not nearly as much of a "blurring" of eastern and western types as found in the flickers was evident yet among the orioles. Indeed, if anything the incidence of hybridization may be decreasing in the Platte Valley, and a larger percentage of parental plumage types seems to be appearing in the combined populations. Thus, perhaps in only a relatively few generations evolution has worked on the population to "weed out" the inferior hybrid types and sharpen up the fitness of the remaining genetic pools by developing methods for reducing further hybridization.

The flickers had remained in the wood lot throughout the mild preceeding winter, although they had been forced to shift from their summer diet of ants to a more typically woodpecker diet of wood-boring insects, which were abundant in the dead and dying elms and old cottonwoods in the area. At times they also resorted to eating hackberries, mulberries, poison ivy, and the fruit or berries of other trees, shrubs, and vines that remained available through the winter months. As spring approached the male began drumming behavior within his territory, and selected a tall, firm branch of one of the trees in the wood lot to drum out his territorial signal. At times he even flew into the farmstead near the wood lot and drummed on some sheet metal that partly covered the roof of the decaying barn. Indeed, he drilled a small hole in one of the farmhouse walls and used the cozy site as a winter roost, while the female continued to roost in her nesting hole of the previous summer. Together with his spring drumming, the male regularly uttered his loud *wicker, wicker, wicker* song, which also served to stake out the bird's claim to this particular wood lot as his own. It was one of the earliest bird songs of spring, and its ringing quality portrayed a sense of optimism and revival in the otherwise seemingly lifeless early March forest surroundings.

The pair's last nesting hole had been drilled near the very top of a cottonwood snag. It consisted of a perfectly round opening drilled into the underside of the tilting limb, thus keeping most rain out of the hole. Although the pair had not remained in contact through the winter months, both returned to their nesting hole of the previous year, and therefore mate recognition was easily attained. But over the winter the hole had become soggy with snow and rain, and the repeated periods of freezing and thawing had made it rather unsuitable for reuse. Therefore the male selected a new site—somewhat lower than the old one, but on the same limb—and began to excavate a new hole. Most of the work was done by the male, with some help from his mate, and when the excavation was nearly complete he began an intensive period of courtship display, calling and spreading his wings and tail to expose their colorful undersides to his mate. Mating occurred about the time the excavation had been completed, and egg laying began almost immediately thereafter. The female laid seven eggs on a daily basis, and incubation began after the last egg was produced. The male participated actively in this behavior, and indeed he did nearly all the nighttime incubation.

It was not until the flickers were already incubating in early May that the first of the orioles returned to the wood lot. Then suddenly they were everywhere, casting little flickerings of orange flame through the leafing trees, and their rich whistled songs penetrated the forest upperstory. The stands of elms and cottonwoods were quickly sought out by the territorial males, for these trees provided a perfect nesting environment. Their great stature, with gracefully sweeping lateral branches and extensive open space below the leafy canopy, offered ideal sites for the hanging nests of the species.

After a period of aggressive interactions and chases, the wood lot was essentially divided up among three pairs of orioles, each of which occupied about a third of the total area and also extended their foraging range well out into the pasturelands around it. As the territorial limits became established, each of the males also acquired a mate, and very quickly the females began nest building. In slightly under a week each had fashioned a remarkable hanging structure, totally unlike the nests of the other birds nesting in the wood lot. The nests were constructed of grasses and other fibrous materials in what seemed at first to be a rather haphazard method that involved simply weaving and pulling together these materials so that a woven structure quickly took shape. The nest was long and stockinglike, somewhat deeper than it was wide, and in most cases it had a somewhat lateral rather than strictly vertical opening. The males participated little if at all in this activity and instead spent most of their nonforaging time maintaining their territorial boundaries. In two of the cases, the females built their nests only a few feet from the remains of the one from the year before, and occasionally they even used the old nests for scavenging building materials. Nearly all the nests were also near those of eastern or western kingbirds, sometimes being situated as close as five or ten feet away. Whether the kingbirds were attracted to the orioles or vice versa was impossible to discern, but quite probably it was the orioles that chose sites near kingbird nests, for no other species of songbird was so alert to possible nest dangers or so quick to respond by screams, threats, or actual attacks. Kingbirds were even more common than orioles in the area, and like the orioles they liked to place their nests in commanding sites close to open pastureland, where they could easily sally forth and capture flying insects. But their nests were much less conspicuous and were usually tightly lodged in the crotches of relatively low and vertically oriented branches, rather than being exposed to the sun and wind at the ends of the branches.

By the time the orioles and kingbirds had begun their nests, the whole wood lot was alive with breeding birds. A small colony of common grackles had begun to nest in a cluster of low cedars, and robins were building their mud-lined nests and placing them almost everywhere, including the windowsills and eaves of the deserted farm buildings and the branches of various trees in the wood lot. House wrens shattered the late spring mornings with their incessantly cheerful songs, and bluejays flitted from tree to tree, always on the lookout for unguarded nests with eggs or nestlings to steal. The insect population of the area was swarming, and there was an abundance of food for all.

An upland sandpiper landing on a perch in a prairie meadow. Photo by author

The Prairie

The male meadowlark hunched over as the snow pelted down, obscuring his vision and causing his territory to turn into a shapeless sea of white before his eyes. He had wintered in the Republican River Valley, and now he had returned and was trying to establish his territory in a new area, for his past-year's territory was in the center of a newly established center-pivot operation that had begun in late spring the year before. As result, the nesting attempt that he and his mate had undertaken then had been destroyed, and they had not been able to acquire a new territory in time to renest successfully. This year, his chosen territory lay just outside the enormous circular swath that the center pivot operation had cut into the native grassland vegetation, and the insects attracted to the irrigated cropland would provide an abundant food supply for him, his mate, and their brood. But the operation had done great damage to the general breeding bird population of the area. Other than a few other meadowlarks and horned larks that had tried to nest in the irrigated cornfield, the only new species to be attracted was the killdeer. The grasshopper sparrows, dickcissels, and nearly thirty other prairie nesters had abandoned the area as soon as it had been converted to cornfields, and now they were either confined to border areas or had disappeared altogether.

Luckily, there were only a few areas that were sufficiently free of tree growth, and not wet enough for corn, to be worthy of consideration for center-pivot irrigation, and only two center pivots had so far been successfully installed. The rest of the prairie areas mantling the island's interior had either been left intact and used for grazing or had been converted to alfalfa hayfields or dryland corn. Those converted to hayfields still supported much the same breeding birds as did native prairies, with large populations of mourning doves, meadowlarks, dickcissels, and cowbirds present. The dryland cornfields were not greatly different from the irrigated ones as far as their limited birdlife was concerned.

The snow-drenched meadowlark revealed by its song that it was of the western species, although to the average observer this was not apparent, for only its longer, more complex, and less slurred song notes made it possible to distinguish the bird from the nearly identical eastern meadowlarks, which were also nesting on the island. By and large, the wet meadows were the areas primarily preferred by the easterns, while the somewhat drier prairie areas were especially favored by the westerns. Although very closely related, and indeed treating each other as if they were one and the same species by expelling all other male meadowlards from their territories, the birds nonetheless only rather rarely hybridized. Evidently females were able to distinguish males readily by their songs and minor call differences and were attracted to their own species, and thus hybridization between the two species was usually avoided.

The early April snowstorm was not a serious setback for the meadowlarks, for most of them had only recently established their territories, and none had yet begun egg laying. Indeed, some had not even acquired mates yet, although this particular male had managed to attract his mate of the last season, and thus little time had been wasted in reforming their pair bond. She was now at work on a new nest. The site was at the base of a clump of little bluestem, which formed a dense canopy overhead, obscuring it to the eyes of avian predators. Furthermore, she had drawn many of the grass stems over the bowl of the nest, forming a complete overhead roof, and it had a lateral entrance. The earthen bowl itself was lined with much finer strands of grama grasses, on which the eggs would be laid. The opening of the nest was toward the east, away from the prevailing spring and summer winds, and the whole structure rather surprisingly resembled a Pawnee earthlodge in its shape, rounded canopy, and eastern orientation. It was situated well inside the male's seven-acre territory, which he advertised many times a day by singing from various song perches or occasionally while in flight.

By mid-April the nest had been completed, and egg laying began. Five white and brown spotted eggs were laid in the nest during the next five days. The female was particularly shy about approaching and leaving her nest during this period, for cowbirds were abundant in the area, and meadowlarks are one of the prime hosts for parastically laid eggs in Nebraska. But there was a more than adequate supply of dickcissels, field sparrows, red-winged blackbirds, and other much more easily found nests for the cowbirds to exploit, and thus the female meadowlark managed to complete her clutch without having it discovered either by cowbirds or any of the other enemies of the species.

A dickcissel territory was becoming established within a hundred yards of the meadowlark nest, in a weedy area that had been heavily grazed in previous years and that now had a substantial number of thistles scattered among the prairie grasses. A male that had arrived in late April established a small territory of slightly less than an acre in size—about the smallest that might be expected to attract a female and successfully support a family. The population density of dickcissels was great, and many males were competing for favorable territories. At least one of the important components of these territories was the presence of broad-leaved herbs, since these are favored locations for nest sites; in their absence the nest is usually placed on the ground. When leafy herbs are present the nests tend to be well concealed from above, perhaps making them less conspicuous to cowbirds and possibly also moderating the daytime nest temperature variations.

The resident dickcissel male had selected a very good territory from these standpoints; beside the thistles there were also heavy growths of asters, sunflowers, and ragweeds, all providing an abundance of shady cover. Indeed, no less than three females became attracted to his territory, and he mated with all three. On the other hand, some of the territorial male dickcissels with holdings as large or larger than his own went without mates, for their territories did not hold the number and quality of potential nest sites that this one did.

It was early May before the first female dickcissel had constructed her nest, which she did without any help from the male. Instead, he tended his small territorial holdings assiduously, combining incessant singing with chases of neighboring males and sexual chases of his prospective mates, which rarely if ever ended in actual copulation. Instead, mating typically occurred at exposed sites, after an invitational display by the female. Nests were completed in only a few days time, and as soon as one female had completed nest building the male would turn his attention to a new bird, courting her just as actively as he had the previous one. By the start of June, all three of the dickcissel females were sitting on clutches. In two cases a cowbird egg was also present, and one of the nests held five cowbird eggs in addition to three dickcissel eggs, virtually filling the entire nest cavity. Perhaps because of the high parasitism level, this female abandoned her clutch shortly after it was completed and began nesting again in another male's territory a few hundred yards away. Another of the nests was blown down in a windstorm about halfway through the incubation period. Only the third, which held four dickcissel eggs and the single cowbird egg, actually hatched.

The cowbird egg hatched nearly a day sooner than did the dickcissels, after only eleven days of incubation. The cowbird egg's spotted markings and lack of a blue background color had easily distinguished it from those of the dickcissel, but the young nestlings were much more similar. The spare down on the cowbird was an olive gray, and that of the dickcissels pure white. However, the baby cowbird had a deep reddish pink gape color, with bright yellow flanges at the bases of the bill. This brilliant 'target," together with the approximate day's headstart in being fed by the female, gave the young cowbird a distinct competitive advantage over its nestmates. Of the four baby dickcissels, two eventually died as a result of competition for the limited food resources, while the other two survived to fledge, along with the cowbird. Even after the young cowbird left the nest it continued to follow the female dickcissel for a time to beg for food, and for nearly a week the dickcissel continued to feed it as well as her two remaining young. Finally they all became independent, and the female began a second round of courtship behavior, in preparation for a second nesting attempt in July.

When the dickcissel nest that had been abandoned was discovered by a female deer mouse one evening in its nocturnal wanderings, it was quickly adopted as a possible nest site. Had there been any trees nearby the mouse would almost certainly have chosen such a nesting site, but none were available, and the well-hidden dickcissel nest needed but little modification to make it suitable for the mouse. First, a dome-shaped roof was added to the nest, leaving only a small side entrance that the animal was able to close over from the inside. Then, feathers and fur were added to the inside to insulate the nest and provide a comfortable, fluffy lining. Finally, various stems and roots of vegetation were worked into the outer framework, disguising the nest and perhaps also adding to its insulation. After it was finished the nest rather resembled that of a harvest mouse, and indeed harvest mice were also present in the area, especially in the wet meadow sites. But this particular mouse was a fawn-colored deer mouse, which had been born the previous fall and had wintered as part of a group of about a dozen animals that had huddled together in a common nest through the coldest weather. With spring, she had moved out on her own after being fertilized and had established a small home range of about a half-acre in size that more or less fitted inside the male dickcissel's territory. The two animals rarely if ever saw one another, for the mouse slept while the dickcissel was active during the day, and the bird was roosting at night when the mouse was out and about searching for much the same sort of foods as those of the bird. These included grasshoppers, seeds of grasses and weeds, grain crops when they were available, and occasionally snails and worms that she chanced to encounter on her nightly forays. Many of the items that the mouse found were eaten on the spot, while others were taken to a small feeding platform on another old dickcissel nest. The mouse would carry a small hoard of food back to this nest in her cheek pouches, empty them, and select some for immediate consumption. The excess would be stored in the nest for later use, when she would be more fully occupied with youngsters and have less time for extensive food hunting.

The female had been fertilized almost three weeks before and even now was nearly ready to give birth. A few nights later she left her nest for the last time to gather food and then returned to await the birth of her litter. It was indeed a fairly good-sized litter, considering that it was her very first pregnancy; four wrinkled, naked, flesh-colored babies emerged, albeit with a bit of help from their mother by gentle prodding and pulling with feet and teeth. Three days after they were born the ears of the young unfolded, and three days later their incisor teeth broke through the gums. By then the young were able to hold on so tightly to the female's nipples that she would drag them along with her for a short distance when she tried to move. But when she wanted to move them any distance, she would pick them up by the belly and form them into a small ball with her front feet, so she could move them around without tripping over their legs or tails. Such moving became necessary about two weeks after they had been born, for the nest had become badly fouled with their droppings. Thus they moved to a new nest site when the youngsters were about two weeks old and becoming well furred, and their eyes were opening. Within another week they were weaned, and by the time the females were nine weeks old they too were ready to mate. The males matured about ten days later than the females, but by midsummer all the young had moved out of the nesting area to begin reproduction themselves. Many of them were taken by coyotes, cats, shrews, and owls as they wandered about tentatively, unaware of the best runways and tunnels to use to try to escape danger. Indeed, the female's mate had been killed by a great horned owl only a short time after fertilizing her, and she too had to go out and locate a new mate before breeding a second time that summer. Thus, it was necessary that the female seek out a new mate, and one evening she wandered beyond her usual limits of foraging, on the lookout for scent marks of a mature male. So intent was she on this particular activity that she didn't catch the scent of shrew in the air, a male short-tailed shrew also on the lookout for a mate. The two animals were using the same runway, and they suddenly blundered into one another. The mouse's better hearing and eyesight were not enough to save her, for with a lightning fast movement the shrew pounced on the mouse, knocking her to the ground. The shrew quickly injected a supply of venom from its salivary glands, which entered the mouse's throat from a wound made by the shrew's teeth. Almost immediately the mouse's heartbeat slowed down and her breathing almost stopped, effectively paralyzing the animal. The shrew pulled the dying mouse back through its subterranean runway for some distance before stopping to devour her, which it did in a surprisingly short time. Two nights later the shrew was itself killed by a coyote, which started to eat it but then quickly spit it out because of its strong musky odor. The carcass was finally found and eaten by an opossum.

A howling coyote in late winter. Photo by author

There at the top of the hills where the skulls lie,
There where the streams of water shorten people's legs,
Where once great herds of buffalo started from,
Yet I saw nothing yonder but an expanse of land.
Only the expanse of land, only the expanse of land.

Pawnee song (Densmore, 1929)

The Future

Great blue herons sometimes nest in tall elms that occur along the Platte. Photo by author

The Future of the River

The Platte has always been a river of extremes. During the early period of westward expansion it presented an impassable barrier to horses and wagons from the vicinity of Fort Kearney eastward, and many animals were swept downstream to their deaths in the annual spring floods before the harnessing of the river. Indeed the Pathfinder Reservoir in eastern Wyoming, the second dam in the nation to be built and completed (in 1909) under the provisions of the Reclamation Act of 1902, was constructed to help control the river. Well before that time, or by about 1870, irrigation use of the Platte's water had begun, and even before the turn of the century there had been a great expansion of irrigation canals and ditches in the Platte Valley.

Accurate information on river width and rates of water flow before 1900 are rather hard to obtain, but the estimates of six explorers of the North Platte and Platte rivers between 1813 and 1857 indicate channel widths ranging from one-half to three miles wide. In 1882 the width of the Platte just below the junction of its north and south branches was measured at 5,350 feet. Average peak flow measurements before April of 1909 at North Platte were more than 500 cubic meters per second (using an average of 14 years). On the other hand, peak flows for the 13 years between 1957 and 1970 averaged only 72 cubic meters per second. Similarly, the width of the river has declined greatly in the past century, especially in the area between Minatare and Overton, where the 1977 channel width was only about ten to twenty percent of that of the 1865 channel. Virtually all these river channel changes and reductions in annual discharges are apparently the result of increased human water consumption and the creation of on-stream reservoirs that divert flow for agricultural purposes or otherwise regulate the seasonal flow of the river.

Within the Platte Valley of Nebraska, over 567,000 acres of land are now being irrigated from surface water sources, and nearly two million acre-feet of water are being diverted by forty canals and twenty-five stream bank pumps. A total of fifteen reservoirs are now present in the channel of the North Platte River or within the Platte Valley near Scottsbluff and at various off-channel locations between Sutherland and Lexington. In the North Platte Valley nearly 1.4 million acre-feet of water are being used to irrigate 325,000 acres, while in the South Platte Valley 22,230 acre-feet irrigate 9,700 acres; and below the confluence of these two branches another 175,000 acres are irrigated with 504,330 acre-feet of water.

Altogether probably about 70 percent of the historical water flow of the Platte system has now been diverted to these various uses in Wyoming, Colorado, and Nebraska; indeed Colorado now has total legal control over the South Platte's water rights from October 15 to April 1. Now only about 20 percent of the water annually reaching the confluence of the North and South Plattes comes from the South Platte system.

As irrigation supplies have dwindled, new clamorings for use of the remaining water in the Platte have intensified. Between 1976 and the end of 1982 at least seven major applications for water diversions were under consideration by the Nebraska Water Resource Commission, including not only intrabasin applications but also some applications for transbasin diversion projects. One would move water into the basin of the Little Blue River and would bring 125,000 acre-feet of water into the Little Blue basin, to irrigate 66,500 acres of land in that drainage basin. Preliminary approval of transbasin diversion by the Nebraska Supreme Court in 1980 stimulated a number of additional proposals for transbasin diversions.

The legal problems associated with all of these requests to use the Platte's water are staggeringly complex, and even if most of them were eventually approved and funded they would more than use up the current annual flow of about 600,000 to 900,000 acre-feet of water now present in the central Platte Valley. Thus, another proposed transbasin diversion into the Big Blue drainage system would irrigate 48,000 acres in Hamilton County at a cost of some 315,000 acre-feet of water annually. Together, these two transbasin projects would use more than half of the Platte's current flow, irrespective of all of the proposed intrabasin projects.

Of more immediate concern to Nebraskans is the proposed joint Twin Valley-Prairie Bend project proposal, which has been advanced by the Central Platte Conservation Association (representing landowners from the Shelton, Wood River, and Cairo areas) and the Twin Valley Conservation Association (representing land interests in the Elm Creek and Kearney areas). This plan would divert water from the Platte at a point between Lexington and Overton, run the water north via a series of canals and small reservoirs, and irrigate farms in Buffalo and Hall counties, with any excess water eventually returning to the Platte via the Wood River or possibly recharging the underground aquifer and ultimately seeping back to the Platte. The potential withdrawal of the Twin Valley portion of the project includes 165,000 acre-feet annually for irrigation and 213,800 acre-feet per year to provide reservoir storage capacity. Additionally, the Prairie Bend portion would remove 210,000 acre-feet annually for irrigation and 352,000 acre-feet for impoundment purposes. The combination would thus have a potential annual withdrawal of 940,800 acre-feet, or virtually the same amount as the average annual total flow reported for the Platte River in the Overton area over a twenty-five-year period, or 930,540 acre-feet.

Nebraskans will therefore soon have to decide whether the entire flow of the Platte River should be handed over to a relatively few farming interests in these counties, thereby destroying the downstream river. Proponents of the project have argued that some seepage will eventually reach the Platte again and thus allow for a certain flow of water. However, the amount of this flow and the length of time required for it to begin are both being disputed, and there is no certainty that any flow will occur at all. If indeed it does occur, the water is likely to be high in nitrates and other pollutants such as pesticides associated with agriculture.

Loss of the Platte River below Overton would have myriad ecological and economic effects. These include detrimental effects on municipal wells located along the Platte, destruction of the subirrigated and wet meadow areas near the river, reduction of river-dependent gravel-pit lake levels (affecting associated residential and recreational development in the area), destruction of fish and wildlife habitat associated with the river, deterioration of water quality in the Platte and its associated aquifers by return seepage of nitrate-rich waters, and reduction of flood control capacity in the lower Platte Valley as a result of vegetational encroachment on previously free-flowing channels.

On the other hand, supporters of the project suggest that, without supplementary water supplies, groundwater levels in the area are likely to decline an additional twenty to eighty feet below predevelopment levels by the year 2020 owing to well pumping for irrigation, thereby causing increased water costs. Moreover, they have noted that reductions of water in the subsurface aquifer will further concentrate already dangerous levels of nitrates that have resulted from intensive fertilization in the area and will also intensify public health concerns as to the safety of municipal water supplies in towns east of Kearney. Since both of these problems are directly the result of previous reckless use of the land's resources, it seems questionable whether a river should be destroyed simply to encourage and subsidize the very resource management practices that initially produced the concerns.

The major purpose of this rapid overview of past and proposed uses of the finite supply of the Platte's water is to point out that the future of the river is going to be dependent on decisions made by a few people as to how the water is to be used, and for whom. Traditional legal uses of the Platte's water, such as municipal water supplies and agricultural and industrial needs, have not included consideration of the historic, aesthetic, recreational, or conservational values of the river. Nor has the state yet been forced to come to grips fully with the question of how it can meet federal requirements for maintaining the critical habitats of endangered species while still trying to meet local and regional pressures concerning agriculturally related problems that affect the economy.

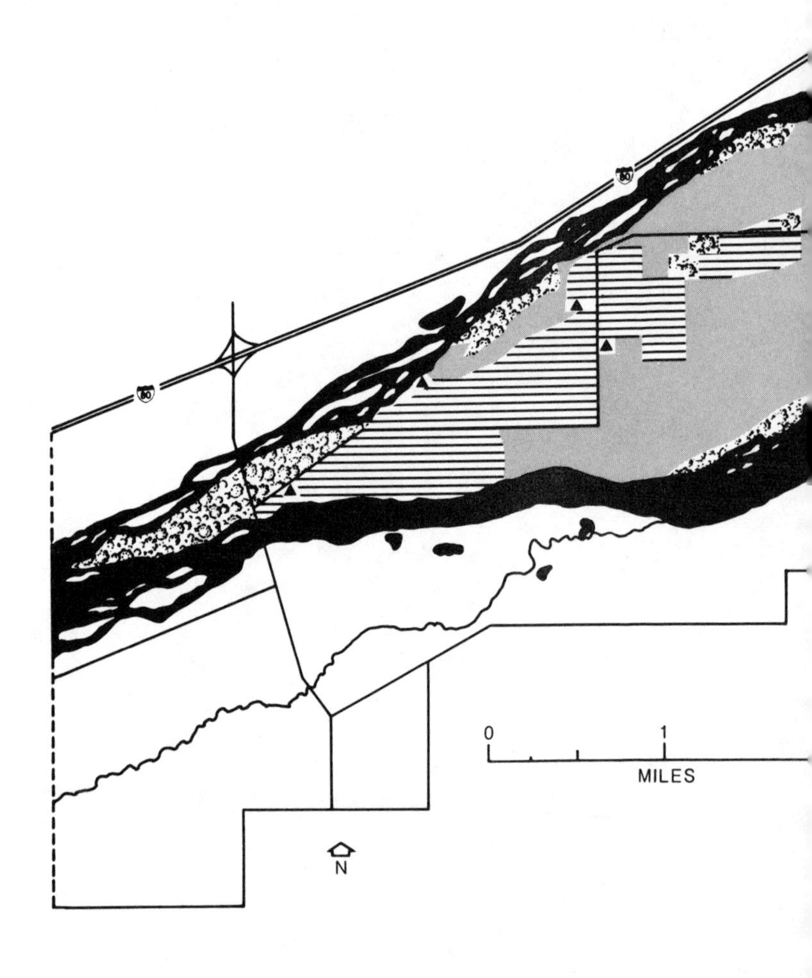
80
80
0
1
MILES
N

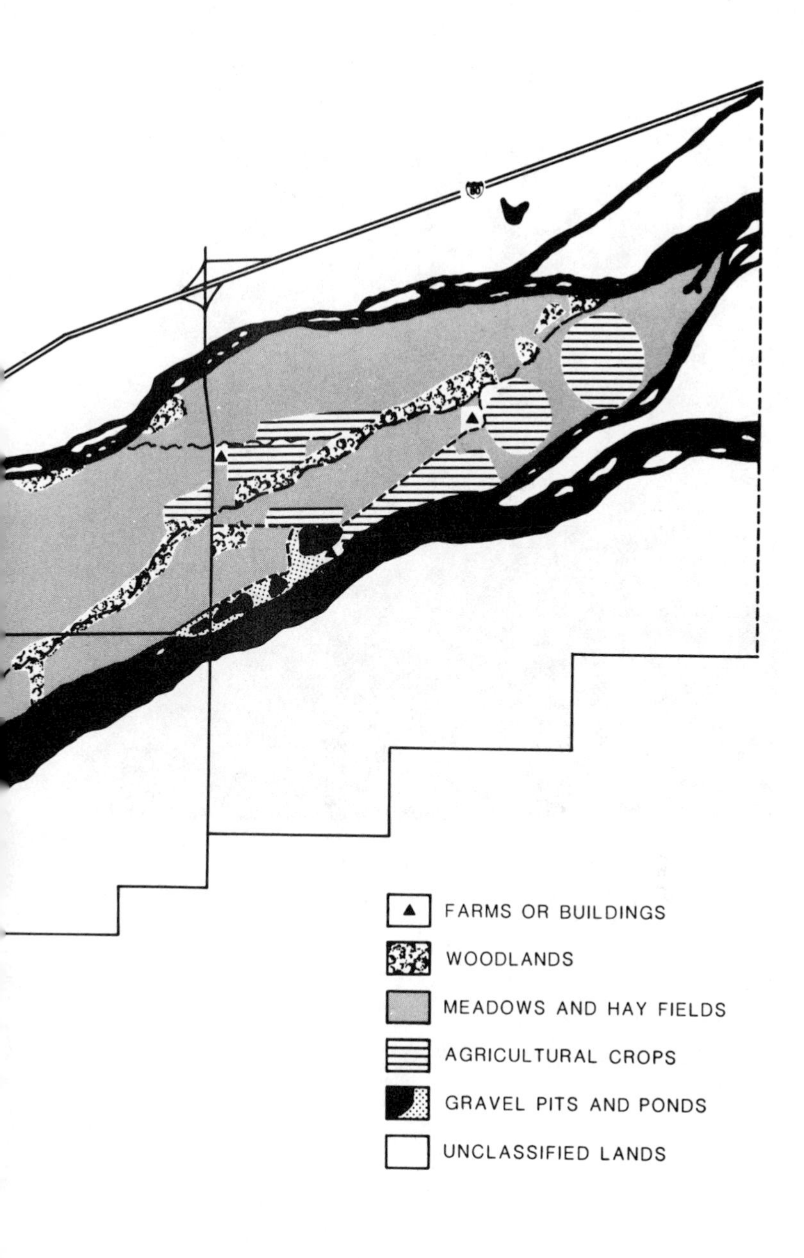

FIGURE 3: Habitat map of Shoemaker Island, based on aerial photography and ground observations in 1982.

An adult bald eagle, a common visitor to the Platte from late fall to early spring. Photo by author

The Future of the People

The past, present, and future of the people of Nebraska are inextricably linked to the Platte River. It was the early irrigation development of the South Platte valley of Colorado in the 1880s that opened up northeastern Colorado to cultivation, and a similar development of irrigation in the North Platte valley of southeastern Wyoming, starting in the 1870s, was largely responsible for the economic development of that area. Similarly, irrigation projects begun in the 1870s in the area around North Platte were to have far-reaching effects on the economy of western Nebraska, in spite of early political reservations about the possible adverse effects on land values that recognition of

Nebraska as an irrigation-dependent state might produce. In spite of these and many other problems, irrigation gradually developed in western Nebraska in the late 1800s and early 1900s. In fact, between 1919 and 1929 there was a ninefold increase in the amount of irrigation by well pumping in the state, which was the greatest such rate of increase in all of the nineteen irrigated western states for that time period.

By 1930 irrigation efforts were extending eastward all the way to the middle Platte Valley. By then, 250 of Buffalo County's 2,429 farms were irrigated, representing a 762 percent increase in irrigation activity during the period 1920 to 1930. Of Hall County's 1,628 farms in 1930, 91 were irrigated. By 1954 there were 1,386 farms in Hall County, representing a reduction of about 250 farms in 24 years, with a corresponding increase in average farm size (228.2 acres in 1954). Between 1939 and 1954 the acreage of Hall County farms planted to corn almost doubled, primarily because of irrigation.

The first irrigation wells were drilled in Hall County in 1912. There were more than 100 wells in use by 1929, and more than 1,000 by 1947. By 1957 there were over 2,000 irrigation wells in Hall County, and nearly 60 percent of the cropland there was then under irrigation. The development and introduction of center-pivot irrigation in the state during the 1950s added a new dimension to the story, for undulating land not previously suited to gravity irrigation could then be watered, albeit at a substantial increased cost in equipment, water usage, and energy consumption. In 1972 there were only five center pivots in all of Hall County. By 1975 this figure had risen to 57; by 1980, the number was 159, or more than a thirtyfold increase in less than a decade. In 1980 there were also only 884 active farms in Hall County, or about half the number present in 1930. Yet, nearly 65 percent of the land (209,000 acres) was under irrigation, compared to less than 5,000 acres being irrigated in 1930. Statewide by 1980 there were probably over seven million acres under irrigation, of which about 30 percent was being watered by center-pivot or other sprinkler irrigation systems. By then, Nebraska ranked third in the nation in terms of total irrigation acreage, behind only California and Texas.

Obviously, the size and style of farming in Nebraska has changed drastically in the past half-century and will continue to change as water supplies, energy costs, and other farming economics dictate.

For better or worse, the Platte Valley and indeed much of the state has become dependent on irrigation to keep farming a profitable enterprise, and a decrease in water supplies through declining water tables or loss of surface water supplies would cause virtual disaster to the economy of the state. Reductions in subsurface levels of underground water supplies associated with the Ogallala aquifer in some areas of the state are already causing some local problems, and the ecological and legal relationships between surface and underground water supplies pose problems of a magnitude that neither scientists nor politicians have been able or willing to address properly. Yet solving such problems is critical to the future of the state and its citizens.

An ancient elm on Mormon Island. Photo by William S. Whitney, Flatland Impressions, Aurora, Nebraska

The Future of the Past

To a large degree, it is senseless to hope that the past glories of the Platte River can forever be preserved, for they are mostly already lost. No living person can remember or even visualize how the Platte must have appeared when it was first seen by the early European explorers, nor can anyone imagine the sounds and sights of a herd of thousands of bison fording the river. Together with the prairie wolves, the Pawnees, the Conestoga wagons, and our view of an unspoiled American wilderness, they are gone forever.

But the Platte is not without its present-day splendors or a unique kind of beauty. A spring sunset, with the sky garlanded by cranes and geese, is a sight to be remembered for a lifetime. A lazy summer morning, the air sweetened by the songs of orioles, grosbeaks, wrens, and doves, and redolent with the scent of plum trees and prairie roses, provides a setting as rich as a Beethoven symphony. A crisp fall morning, with golden cottonwoods outlined against an azure sky and a brisk north wind warning of impending winter, is a testament to the beauty of life even as it is fading. And a winter's day, with only the sound of creaking ice and drifting snow to disturb the silence, provides a time to ponder hopes for the inevitable spring. For as certainly as the ice will break up and the flood waters of the upper Platte will bring down with them new lifegiving waters from the invisible mountains far to the west, so too there is a certainty that amid the starkness of the snow-draped landscape there are plants germinating, ground squirrels hibernating, and sandhill cranes massing in the warmer lands to the south, waiting only for the proper moment to return again to their ancestral meeting grounds of the Platte Valley.

These present-day wonders of the Platte are thus all the more to be guarded and cherished, for they represent a tangible legacy from Nebraska's past. And the sentient soul that visits the Platte, if only in the imagination, can rest under the same old cottonwood tree that may have once offered its shade to a tired immigrant, can hear the water's endless conversation with itself, telling the same stories it told a century earlier, and can watch the sun dip below the same western horizon that once beckoned American dreamers and eventually became a large part of the American dream itself. In so doing, it is hard not to believe that, while the value of water might be measured by the gallon or acre-foot, the values of a river are best measured by the fact that its waters are the gifts of rains and snows in times past and places far removed, and the most sensible and generous thing we can do is to let it flow ever onward, to grace another place and someone else's future.

Behold! Our Mother Earth is lying here.
Behold! She giveth of her fruitfulness.
Truly, her power gives she us.
Give thanks to Mother Earth who lieth here.

Behold on Mother Earth the growing fields;
Behold the promise of her fruitfulness!
Truly, her power gives she us.
Give thanks to Mother Earth who lieth here.

Behold on Mother Earth the spreading trees!
Behold the promise of her fruitfulness!
Truly, her power gives she us.
Give thanks to Mother Earth who lieth here.

Behold on Mother Earth the running streams!
Behold the promise of her fruitfulness!
Truly, her power gives she us.
Give thanks to Mother Earth who lieth here.

From Pawnee Hako ceremony
(Fletcher, 1900–1901)

Appendix

Great egrets are rare summer visitors to the Platte valley. Photo by author

Checklist of Birds of the Middle Platte Valley[1]

Eared Grebe (*Podiceps nigricollis*). An uncommon migrant in the Platte Valley; one spring (May) record for Mormon Island. Sandpits are typical habitats.

Pied-billed Grebe (*Podilymbus podiceps*). A common migrant in the Platte Valley; migrant and summer visitor on Mormon Island. Ephemeral ponds and sandpits are typical habitats.

American White Pelican (*Pelecanus erythrorhynchus*). Uncommon

1. Mormon Island records are primarily from Lingle and Hay (1982 and personal communication)

migrant along the Platte; reported at Mormon Island only during April and May. Associated with river channels.

Double-crested Cormorant (*Phalacrocorax auritus*). Uncommon migrant along the Platte; reported at Mormon Island only during April and May. River channels and sandpits are typical habitats.

Great Egret (*Casmerodius albus*). Occasional migrant and summer visitor in the Platte Valley; reported only once (July) at Mormon Island. The typical habitat is river shorelines.

Great Blue Heron (*Ardea herodias*). Common migrant and summer visitor in the Platte Valley; breeding near Gothenberg. Reported at Mormon Island from April through mid-October. Habitats include river shorelines, islands, and sandbars.

Green-backed Heron (*Butorides virescens*). Uncommon migrant and summer visitor in the Platte Valley, probably breeding locally. Reported at Mormon Island between mid-May and early September. The typical habitat is river shorelines.

Little Blue Heron (*Florida caerulea*). Occasional migrant in the Platte Valley; reported twice (April and May) at Mormon Island. The typical habitat is river shorelines.

Black-crowned Night Heron (*Nycticorax nycticorax*). Common migrant and local summer resident in the Platte Valley; recorded during spring (April) at Mormon Island. The typical habitat is river shorelines.

American Bittern (*Botaurus lentiginosus*). Common migrant and local summer resident in the Platte Valley; recorded during spring (April) at Mormon Island. The typical habitat is sandpits or marshy areas.

Canada Goose (*Branta canadensis*). An abundant migrant and local breeder in the Platte Valley. At Mormon Island, abundant in spring, a local summer resident, common in fall; and an occasional overwintering species. Typical habitats are river channels for roosting and wetland meadows or croplands for foraging.

Greater White-fronted Goose (*Anser albifrons*). An abundant migrant in the Platte Valley; at Mormon Island, abundant in spring, rare in fall, and an occasional overwintering species. Typical habitats are river channels for roosting and wetland meadows for foraging.

Snow Goose, including "Blue Goose" (*Anser caerulescens*). An uncommon migrant in the Platte Valley; at Mormon Island uncommon in spring and occasional in late winter. Typical habitats are river channels for roosting and wetland meadows or croplands for feeding.

Mallard (*Anas platyrhynchos*). An abundant migrant and local summer resident in the Platte Valley; at Mormon Island abundant in spring, uncommon during summer, and fairly common during fall, occasionally overwintering. Known to breed at Mormon Island. Typical habitats are river channels, sandbars, wetland meadows, and corn fields.

Gadwall (*Anas strepera*). A common migrant and summer resident in the Platte Valley; at Mormon Island uncommon in spring and rare in fall, with records from April to late September. Typical habitats are river channels.

Northern Pintail (*Anas acuta*). An abundant migrant and local summer resident in the Platte Valley; at Mormon Island abundant in later winter and spring, uncommon in summer, and uncommon in fall, sometimes overwintering. A potential breeder at Mormon Island. Typical habitats are river channels, wetland meadows, and corn fields.

Green-winged Teal (*Anas crecca*). An abundant migrant and probable local summer resident in the Platte Valley; at Mormon Island common in spring and fall, and rare during summer and winter. A potential breeder at Mormon Island. Typical habitats are river channels and wetland meadows.

Blue-winged Teal (*Anas discors*). An abundant migrant and common summer resident in the Platte Valley; at Mormon Island a common migrant and uncommon summer resident, known to breed. Typical habitats are river channels, wetland meadows, and sandpits.

American Wigeon (*Anas americana*). A common to abundant migrant in the Platte Valley; at Mormon Island a common spring migrant and an uncommon late winter visitor, reported from mid-February to May during spring, no specific fall dates. Habitats include river channels and wet grassy meadows.

Northern Shoveler (*Anas clypeata*). A common to abundant migrant in the Platte Valley, and a local summer resident. At Mormon Island a fairly common spring migrant, a rare summer resident, and fall migrant, reported from mid-March to late September. Typical habitats include river channels and wetland meadows.

Wood Duck (*Aix sponsa*). An uncommon migrant and summer resident in the Platte Valley, decreasing to the west. At Mormon Island an uncommon spring migrant, a rare summer resident, and a potential breeder. Reported as early as mid-March. Typical habitats include riparian woods, sandpits, and wetland meadows.

Redhead (*Aythya americana*). A common migrant in the Platte Valley; at Mormon Island a rare late winter visitor and spring migrant. Reported from late February until April. Typical habitats are river channels and sandpits.

Canvasback (*Aythya valisineria*). An uncommon migrant in the Platte Valley; at Mormon Island a rare spring migrant. Reported from mid-March to late April. Typical habitats are sandpits.

Greater Scaup (*Aythya marila*). A rare migrant in the Platte Valley, but reported on the U.S. Fish and Wildlife Services list of species identified in the Platte River Ecology summary report (1981). Not listed for Mormon Island by Lingle and Hay. The typical habitats are lakes, deep sandpits, and large rivers.

Lesser Scaup (*Aythya affinis*). Common migrant in the Platte Valley; infrequently seen on the Platte River. Not reported for Mormon Island by Lingle and Hay, but common in sandpits of the general area from March to May and from October to late November. Typical habitats are sandpits.

Common Goldeneye (*Bucephala americana*). A common to uncommon migrant in the Platte Valley; at Mormon Island a rare late winter visitor and spring migrant. Reported from February until late April. Typical habitats include river channels and sandpits.

Bufflehead (*Bucephala albeola*). A common to uncommon migrant in the Platte Valley; at Mormon Island a rare spring migrant. Reported in March and April. Typical habitats are sandpits.

Common Eider (*Somateria mollissima*). Accidental in Nebraska, but included in the U.S. Fish and Wildlife Service's list of birds identified in the Platte River Ecology Study summary report.

Hooded Merganser (*Lophodytes cucullatus*). An uncommon to occasional migrant in the eastern Platte Valley, rare farther west. At Mormon Island a rare fall migrant. Reported in November. Typical habitats are sandpits.

Common Merganser (*Mergus merganser*). A common migrant in the Platte Valley, occasionally overwintering. At Mormon Island reported from early March until late April. Typical habitats are river channels.

Turkey Vulture (*Cathartes aura*). An uncommon migrant in the Platte Valley, possibly breeding. At Mormon Island a rare spring migrant. Typical foraging habitats are open fields.

Northern Goshawk (*Accipiter gentilis*). An uncommon migrant in the Platte Valley; at Mormon Island a rare migrant. Reported in April, September, and November. Typical habitats are riparian woodlands.

Sharp-shinned Hawk (*Accipiter striatus*). An uncommon migrant in the Platte Valley; in Mormon Island a rare spring migrant. Reported in March. Typical habitats are riparian woodlands.

Cooper's Hawk (*Accipiter cooperii*). An uncommon migrant in the Platte Valley; at Mormon Island a rare migrant. Reported in April, September, and November. Typical habitats are riparian woodlands.

Red-tailed Hawk (*Buteo jamaicensis*). An uncommon permanent resident in the Platte Valley, breeding locally. At Mormon Island a common permanent resident, nesting in riparian woodlands. Typical habitats are woodlands and wood lots, with foraging in wetland meadows and hay fields.

Swainson's Hawk (*Buteo swainsoni*). An uncommon migrant and summer resident in the Platte Valley, decreasing farther east. At Mormon Island a rare migrant. Observed in April and May. Typical foraging habitats are open fields.

Rough-legged Hawk (*Buteo lagopus*). An uncommon migrant and winter visitor in the Platte Valley, decreasing in the east. At Mormon Island a common winter visitor and variably common migrant from October to March. Typical habitats include wetland meadows, fencerows, riparian woodlands, hay fields, woodlands, idle lands, and croplands.

Ferruginous Hawk (*Buteo regalis*). An uncommon to rare migrant in the Platte Valley; at Mormon Island a rare migrant. Reported during October. Typical habitats are wetland meadows.

Golden Eagle (*Aquila chrysaetos*). An uncommon migrant and winter visitor in the Platte Valley, decreasing to the east. At Mormon Island a rare spring migrant. Reported during March. Typical habitats are wetland meadows and occasionally riparian woodlands.

Bald Eagle (*Halieetus leucocephalus*). An uncommon migrant and winter visitor in the Platte; locally common. At Mormon Island a common winter visitor and early spring migrant. Reported from December to late March. Typical habitats are river channels, riparian woodlands, wetland meadows, hay fields, and shrublands.

Northern Harrier (*Circus cyaneus*). A common migrant and local permanent resident in the Platte Valley; at Mormon Island an occasional winter visitor and spring migrant. Reports are from November to late April. Typical habitats include wetland meadows, hay fields, croplands, idle fields, and fencerows.

Osprey (*Pandion haliaetus*). An uncommon to occasional migrant in the Platte Valley; at Mormon Island a rare migrant. Reported in April and October. Typical habitats are river channels.

Prairie Falcon (*Falco mexicana*). An occasional to rare permanent resident in the western Platte Valley, decreasing to the east. At Mormon Island an uncommon winter visitor and rare migrant, reported from November to March. Typical habitats include wetland meadows and hay fields.

Peregrine Falcon (*Falco peregrinus*). A rare migrant in the Platte Valley; at Mormon Island a rare migrant, reported only during October. Typical habitats include open fields.

Merlin (*Falco columbarius*). An uncommon to rare migrant in the Platte Valley; at Mormon Island a rare migrant. Reported only during March. Typical habitats include riparian woodlands.

American Kestrel (*Falco sparverius*). A common permanent resident in the Platte Valley; at Mormon Island a common resident and potential breeder. Typical habitats include wetland meadows, woodlots, fencerows, croplands, farmsteads, and riparian woodlands.

Greater Prairie Chicken (*Tympanuchus cupido*). A local uncommon resident in the Platte Valley, especially along the eastern Sandhills. Reported during winter at Mormon Island. Typical habitats include tall prairies, wetland meadows, corn fields, and (in winter) riparian woodlands.

Sharp-tailed Grouse (*Pedioecetes phasianellus*). A common resident in the Platte Valley, especially in natural grasslands of the Sandhills. Not reported at Mormon Island. Typical habitats include short- and medium-grass prairies, wetland meadows, corn fields, and (in winter) woody thickets.

Northern Bobwhite (*Colinus virginianus*). A common permanent resident in the Platte Valley; at Mormon Island a common resident and breeder. Typical habitats include woodlots, fencerows, riparian woodlands, shrublands, wetland meadows, and farmsteads.

Ring-necked Pheasant (*Phasianus colchicus*). A common permanent resident in the Platte Valley; at Mormon Island a common resident and breeder. Typical habitats include wetland meadows, fencerows, croplands, riparian woodlands, and brushlands.

Wild Turkey (*Meleagris gallopavo*). A locally common reintroduced permanent resident in the Platte Valley (east to Platte County). Not reported at Mormon Island. Typical habitats include riparian woodlands.

Sandhill Crane (*Grus canadensis*). An abundant spring migrant and uncommon fall migrant in the Platte Valley; at Mormon Island also abundant in spring and uncommon in fall. Reported from early March to mid-April, and from late October to mid-November. Typical roosting habitats are river sandbars and channels; wetland meadows and croplands are foraging habitats.

Whooping Crane (*Grus americana*). A rare migrant in the Platte Valley, previously common in Buffalo and Kearney counties, primarily between Lexington and Grand Island. Recorded in spring from early March to late May, and in fall from mid-September to early November. Not reported in recent years from Shoemaker or Mormon islands, but observed on adjacent wetland areas.

Sora (*Porzana carolina*). A common migrant and local summer resident in the Platte Valley; at Mormon Island a rare spring migrant and breeder. Reported from early May through June. Typical habitats include river channels, wetlands, hay fields, and alfalfa.

American Coot (*Fulica americana*). A common to abundant migrant and summer resident in the Platte Valley, occasionally overwintering. At Mormon Island a very rare spring migrant and summer visitor. Recorded from mid-April through late May. Typical habitats are sandpits and wetland meadows.

Semipalmated Plover (*Charadrius semipalmatus*). An uncommon to occasional spring and fall migrant in the Platte Valley; at Mormon

Island an uncommon spring migrant. Reported from mid-April to mid-May. Typical habitats are sandy river shorelines.

Piping Plover (*Charadrius melodus*). An occasional to rare migrant and summer resident in the Platte Valley; at Mormon Island a fairly common late spring migrant and breeder. Reported from late April to late June. Typical habitats are barren sandbars.

Killdeer (*Charadrius vociferus*). A common to abundant migrant and summer resident in the Platte Valley; at Mormon Island a common migrant and breeder. Reported from early March until late November. Typical habitats are sandbars, wetland meadows, and hay fields.

Lesser Golden Plover (*Charadrius dominica*). An uncommon to occasional migrant in the eastern Platte Valley, decreasing to the west. At Mormon Island a rare spring migrant, reported only in May. Typical habitats are newly seeded fields.

American Woodcock (*Philohela minor*). A rare to occasional migrant in the Platte Valley west to about Grand Island, and a rare and local summer resident as far west as Hall County. Not reported for Mormon Island. Typical habitats are riparian woodlands with deep, moist soils.

Ruddy Turnstone (*Arenaria interpres*). An occasional to rare migrant in the eastern Platte Valley, rarer to the west. At Mormon Island an accidental spring migrant, reported only during May. Typical habitats are river shorelines.

Common Snipe (*Gallinago gallinago*). A common migrant and local or rare summer resident in the Platte Valley; at Mormon Island a common spring and rare fall migrant. Reported from late March to late April, and in September. Typical habitats are wetland meadows and river shorelines.

Long-billed Curlew (*Numenius americanus*). An uncommon migrant in the Platte Valley, and a local summer resident in nearby grasslands. Not reported from Mormon Island. Typical habitats are Sandhills grasslands, meadows, and short-grass plains.

Upland Sandpiper (*Bartramia longicauda*). An uncommon migrant and local summer resident in the Platte Valley. At Mormon Island an abundant spring migrant and breeder. Reported from mid-April through August. Typical habitats are wetland meadows, hay fields, alfalfa fields, and river shorelines.

Spotted Sandpiper (*Actitus macularia*). A common migrant and summer resident in the Platte Valley; at Mormon Island a common migrant and known breeder. Reported from mid-April to late September. Typical habitats are river shorelines.

Solitary Sandpiper (*Tringa solitaria*). A common to occasional migrant in the Platte Valley; at Mormon Island a rare spring migrant. Reported only during April. Typical habitats are wetland meadows.

Greater Yellowlegs (*Tringa melanoleuca*). A common spring and fall migrant in the Platte Valley; at Mormon Island an uncommon spring and fall migrant. Reported from late March to mid-May, and from late July to early October. Typical habitats are river shorelines.

Lesser Yellowlegs (*Tringa flavipes*). A common migrant in the Platte Valley; at Mormon Island a common spring migrant, but unreported during fall. Records are from early April to late May. Typical habitats include wetland meadows and river shorelines.

Willet (*Catoptrophorus semipalmatus*). A common to uncommon migrant in the Platte Valley. At Mormon Island a rare spring and early fall migrant. Reported during April and July. Typical habitats are river shorelines and wetland meadows.

Pectoral Sandpiper (*Calidris melanotis*). A common to abundant migrant in the Platte Valley. At Mormon Island an uncommon spring migrant; not reported during fall. Reported during April and May. Typical habitats include wetland meadows and river shorelines.

White-rumped Sandpiper (*Calidris fuscicollis*). An uncommon spring migrant and apparently rare fall migrant in the Platte Valley; at Mormon Island an abundant spring migrant. Reported from mid-May to early June. Typical habitats include river shorelines and wetland meadows.

Baird's Sandpiper (*Calidris bairdii*). A common to abundant migrant in the Platte Valley; at Mormon Island an abundant spring migrant and occasional early summer straggler. Reported from early April until June 1. Typical habitats include river shorelines and wetland meadows.

Least Sandpiper (*Calidris minutilla*). A common migrant in the Platte Valley; at Mormon Island an abundant spring migrant, not reported during fall. Reported from late April to late May. Typical habitats include river shorelines and wetland meadows.

Dunlin (*Calidris alpina*). An occasional spring migrant in the eastern Platte Valley, becoming rarer to the west, and a rare fall migrant. At Mormon Island an occasional spring migrant, reported only in May. Typical habitats include river shorelines.

Semipalmated Sandpiper (*Calidris pusilla*). An abundant to uncommon migrant in the Platte Valley, less common to the west. At Mormon Island a common spring migrant and uncommon fall migrant. Reported from mid-April to late May and during September. Typical habitats include river shorelines.

Western Sandpiper (*Calidris mauri*). A rare migrant in the eastern Platte Valley, becoming more common to the west. At Mormon Island a rare spring migrant. Reported only in May. Typical habitats include river shorelines.

Short-billed Dowitcher (*Limnodromus griseus*). An occasional to rare migrant in the Platte Valley; at Mormon Island a rare spring migrant. Reported only during May. Typical habitats include river shorelines.

Long-billed Dowitcher (*Limnodromus scolopaceus*). A common migrant in the Platte Valley; at Mormon Island a rare spring migrant. Reported only during April. Typical habitats include river shorelines.

Stilt Sandpiper (*Micropalama himantopus*). A common to uncommon migrant in the Platte Valley, becoming less common to the west. At Mormon Island a rare spring migrant. Reported only during May. Typical habitats include wetland meadows and river shorelines.

Marbled Godwit (*Limosa fedoa*). An uncommon to locally common migrant in the Platte Valley, less frequent during fall. At Mormon Island a rare spring migrant, recorded only during April. Typical habitats are wetland meadows.

Hudsonian Godwit (*Limosa haemastica*). An uncommon migrant in the eastern Platte Valley, rarer to the west. At Mormon Island a rare spring migrant, recorded only during May. Typical habitats are wetland meadows.

American Avocet (*Recurvirostra americana*). An uncommon to rare migrant in the Platte Valley, most common in central areas. At Mormon Island a rare spring migrant, recorded only during May. Typical habitats are alkaline ponds and marshes.

Wilson's Phalarope (*Steganopus tricolor*). A common to abundant migrant and local summer resident in the Platte Valley; at Mormon Island an uncommon migrant and potential breeder. Reported from early May until late June, possibly breeding in native hay. Typical habitats are river channels and wetland meadows.

Herring Gull (*Larus argentatus*). An uncommon migrant in the Platte Valley; at Mormon Island an accidental fall migrant and winter visitor. Records are for October and February. Typical habitats are river channels.

Ring-billed Gull (*Larus delawarensis*). A common migrant and occasional summer or winter visitor in the Platte Valley; at Mormon Island an uncommon migrant and occasional winter visitor. Recorded from November through late March. Typical habitats are river channels.

Franklin's Gull (*Larus pipixcan*). An abundant migrant in the Platte Valley; at Mormon Island an uncommon spring migrant, not reported during fall. Recorded from late March to late May. Typical habitats are river channels and cultivated fields.

Forster's Tern (*Sterna forsteri*). A common migrant in the Platte Valley; at Mormon Island a fairly common spring migrant, not reported during fall. Recorded from late April to mid-May. Typical habitats are river channels.

Least Tern (*Sterna antillarum*). An uncommon and rare summer resident in the Platte Valley; at Mormon Island a rare migrant and occasional breeding summer resident. Reported from mid-May to late July. Typical habitats are river channels with barren sandbars or islands.

Black Tern (*Chlidonias niger*). An uncommon to common migrant in the Platte Valley; at Mormon Island an uncommon spring migrant, not reported during fall. Recorded only during May. Typical habitats are river channels.

Rock Dove (*Columba livia*). An introduced permanent resident in the Platte Valley, occurring near human habitations. Typical habitats are corn fields and farmsteads.

Mourning Dove (*Zenaida macroura*). An abundant migrant and summer resident in the Platte Valley, overwintering in some years. At Mormon Island an abundant migrant and breeding summer resident, common also in fall. Typical habitats include wetland meadows, fencerows, wood lots, croplands, and hayfields.

Yellow-billed Cuckoo (*Coccyzus americanus*). A common migrant and summer resident in the Platte Valley; at Mormon Island an uncommon migrant and summer resident, and a probable breeder; seen rarely during fall. Recorded from late May to early September. Typical habitats are shrublands, fencerows, and farmsteads.

Black-billed Cuckoo (*Coccyzus erythropthalmus*). An uncommon migrant and summer resident in the Platte Valley; at Mormon Island a rare migrant and summer resident, and a probable breeder. Recorded only during May and June. Typical habitats are riparian woodlands, shrublands, and wood lots.

Common Barn Owl (*Tyto alba*). An uncommon permanent resident in the Platte Valley; not included in Lingle and Hay's list for Mormon Island, but listed in the U.S. Fish and Wildlife Service's summary report of the Platte River Ecology Study. Typical habitats are open areas with old buildings, hollow trees, and natural cavities available for nesting.

Eastern Screech Owl (*Otus asio*). A common permanent resident in the Platte Valley; not yet reported from Mormon Island. Typical habitats are riparian woodlands and wood lots.

Great Horned Owl (*Bubo virginianus*). An uncommon permanent resident in the Platte Valley; at Mormon Island an uncommon resident and breeder. Typical habitats include wood lots and riparian woodlands.

Short-eared Owl (*Asio flammeus*). An uncommon permanent resident in the Platte Valley; not reported for Mormon Island but included in the U.S. Fish and Wildlife Services Platte River study summary. Typical habitats are meadows and grasslands.

Burrowing Owl (*Athene cunicularia*). An uncommon migrant and summer resident in the Platte Valley, especially to the west; not reported from Mormon Island. Normally present from late March to mid-September. Typical habitats are grasslands with prairie dog colonies.

Common Nighthawk (*Chordeiles minor*). An abundant migrant and summer resident in the Platte Valley, especially around cities; at Mormon Island a rare to uncommon migrant and rare summer visitor. Records are for May and June. Typical habitats include open areas for aerial foraging and roosts in trees or on posts.

Chimney Swift (*Chaetura pelagica*). A common migrant and summer resident in the Platte Valley; at Mormon Island a rare spring migrant, reported only during May. Typical habitats are open areas for aerial foraging, and chimneys or hollow trees for nesting.

Belted Kingfisher (*Megaceryle alcyon*). A common migrant and summer resident in the Platte Valley, frequently overwintering. At Mormon Island an uncommon permanent resident and breeder. Typical habitats are river channels, riparian woodland edges, sandpits, and shrublands.

Northern Flicker (*Colaptes auratus*). A common permanent resident in the Platte Valley, including Mormon Island. Typical habitats include wetland meadows, riparian woodlands, shrublands, wood lots, fencerows, and croplands.

Red-bellied Woodpecker (*Melanerpes carolinus*). A common permanent resident in the eastern Platte Valley, declining to the west. At Mormon Island an uncommon fall migrant and winter visitor (Lingle and Hay), but reported as common in the Mormon Island area by Short (*Nebraska Bird Review* 29:6). Typical habitats are wood lots, riparian woodlands, and shrublands.

Red-headed Woodpecker (*Melanerpes erythrocephalus*). A common migrant and summer resident in the Platte Valley; at Mormon Island a common to uncommon migrant and common summer resident, presumably breeding. Records extend from late April to late September. Typical habitats are wood lots, riparian woodlands, fencerows, and corn fields.

Hairy Woodpecker (*Piciodes villosus*). A common permanent resident in the Platte Valley; at Mormon Island an uncommon permanent resident and known breeder. Typical habitats include wood lots and riparian woodlands.

Downy Woodpecker (*Picoides pubescens*). A common permanent resident in the Platte Valley; at Mormon Island an uncommon permanent resident and probable breeder. Typical habitats include wood lots, riparian woodlands, and idle fields.

Eastern Kingbird (*Tyrannus tyrannus*). A common migrant and summer resident in the Platte Valley, including Mormon Island.

Records extend from early May to early September. Typical habitats include fencerows, riparian shrublands, woodlands, farmsteads, and wetland meadows.

Western Kingbird (*Tyrannus verticalis*). A common migrant and summer resident in the Platte Valley; at Mormon Island an uncommon migrant and summer resident, presumably breeding. Records extend from early May until mid-September. Typical habitats include wetland meadows, fencerows, shrublands, and newly seeded alfalfa.

Great Crested Flycatcher (*Myiarchus crinitus*). A common migrant and summer resident in the eastern Platte Valley, declining to the west. At Mornon Island an uncommon spring migrant and summer resident, potentially breeding. Records extend from mid-May through late July. Typical habitats include wood lots, riparian woodlands, and fencerows.

Eastern Phoebe (*Sayornis phoebe*). A common migrant and summer resident in the eastern Platte Valley, declining to the west. Not reported for Mormon Island by Lingle and Hay, but considered uncommon by Short (*Nebraska Bird Review* 29:8). Typical habitats are riparian woodlands.

Say's Phoebe (*Sayornis saya*). A common migrant and summer resident in the western Platte Valley, declining to the east. Not reported for Mormon Island by Lingle and Hay, but breeding extends east locally to York County. Typical habitats are open and dry woodlands.

Willow Flycatcher (*Empidonax traillii*). A common migrant and uncommon to rare summer resident in the Platte Valley; at Mormon Island an uncommon migrant and summer resident, potentially breeding. Records extend from late May until mid-September. Typical habitats include riparian woodlands and shrublands.

Least Flycatcher (*Empidonax minimus*). A common migrant in the Platte Valley; at Mormon Island an uncommon spring migrant, not reported during fall. Records are for May only. Typical habitats include riparian woodlands and shrublands.

Eastern Wood Pewee (*Contopus virens*). A common migrant and summer resident in the eastern Platte Valley, declining to the west. Not reported at Mormon Island by Lingle and Hay, but considered common by Short (*Nebraska Bird Review* 29:8) and included in the U.S. Fish and Wildlife Service species identified in the Platte River Ecology Study. Typical habitats are woodlands.

Western Wood Pewee (*Contopus sordidulus*). An uncommon migrant and summer resident in the western Platte Valley, rare or absent to the east. Not reported for Mormon Island by Lingle and Hay, but included in the U.S. Fish and Wildlife Service's list of species identified in the Platte River Ecology Study. Typical habitats are woodlands.

Olive-sided Flycatcher (*Nuttallornis borealis*). An uncommon to occasional migrant in the Platte Valley; a rare spring migrant at Mormon Island, reported only during May. Typical habitats are riparian woodlands and fencerows.

Horned Lark (*Eremophila alpestris*). A common to abundant winter visitor, migrant, and summer resident in the Platte Valley; at Mormon Island an uncommon migrant and winter visitor. Typical habitats include wetland meadows, roadsides, and farmsteads.

Tree Swallow (*Iridoprocne bicolor*). A common migrant and summer resident (possibly west to Hall County) in the eastern Platte Valley, decreasing to the west. At Mormon Island an uncommon spring migrant, reported only during April. Typical habitats are open areas over land or water.

Bank Swallow (*Riparia riparia*). A common migrant and summer resident in the Platte Valley; at Mormon Island an uncommon spring migrant. Records extend from late April to mid-May. Typical habitats are open areas over water or land.

Northern Rough-winged Swallow (*Stelgidopteryx ruficollis*). A common migrant and summer resident in the Platte Valley; at Mormon Island a common migrant and uncommon breeding summer resident. Records extend from early May through June. Typical habitats are riverside banks and open areas above land or water.

Barn Swallow (*Hirundo rustica*). A common migrant and summer resident in the Platte Valley; at Mormon Island a common migrant and breeding summer resident. Records extend from late April until late September. Typical habitats include farmsteads and open areas above land or water.

Cliff Swallow (*Petrochelidon pyrrhonota*). A common to abundant migrant and summer resident in the Platte Valley; at Mormon Island an uncommon migrant and local breeding summer resident.

Records extend from mid-May through September. Typical habitats include river bridges and open areas over land or water.

Blue Jay (*Cyanocitta cristata*). A common permanent resident in the Platte Valley; at Mormon Island variably common throughout the year, and a known breeder. Typical habitats include wetland meadows, riparian woodlands, wood lots, shrublands, and hay fields.

Black-billed Magpie (*Pica pica*). A common permanent resident in the western Platte Valley, declining to the east; at Mormon Island an uncommon permanent resident and known breeder. Typical habitats include riparian woodlands, shrublands, fencerows, and river channels.

American Crow (*Corvus brachyrhynchus*). A common to abundant migrant and common summer resident in the Platte Valley; frequently overwintering. At Mormon Island an uncommon permanent resident and known breeder. Typical habitats include riparian woodlands, river channels, wood lots, wetland meadows, shrublands, and croplands.

Black-capped Chickadee (*Parus atricapillus*). A common permanent resident in the Platte Valley, including Mormon Island. Typical habitats include wood lots, riparian woodlands, idle lands, fencerows, and shrublands.

Tufted Titmouse (*Parus bicolor*). A common permanent resident in the eastern Platte Valley, declining to the west. Not reported for Mormon Island by Lingle and Hay, but considered common by Short (*Nebraska Bird Review* 29:8). Typical habitats are mature riparian woodlands and wood lots.

White-breasted Nuthatch (*Sitta carolinensis*). An uncommon permanent resident in the Platte Valley west to Lincoln County, including Mormon Island. Typical habitats include riparian woodlands and wood lots.

Brown Creeper (*Certhia americana*). A common winter visitor in the Platte Valley; not reported for Mormon Island by Lingle and Hay, but observed on Shoemaker Island. Typical habitats are mature riparian woodlands.

House Wren (*Troglodytes aedon*). A common to abundant migrant and summer resident in the Platte Valley; at Mormon Island a common migrant and a probable breeder. Reported from mid-May to late September. Typical habitats include wood lots, riparian woodlands, shrublands, and fencerows.

Marsh Wren (*Cistothorus palustris*). A common migrant and local summer resident in the Platte Valley (mainly to the north of the river); not reported for Mormon Island. Typical habitats are marshes and wet meadows.

Bewick's Wren (*Thryomanes bewickii*). A rare migrant and local summer resident in the eastern Platte Valley; not reported by Lingle and Hay for Mormon Island. However, it is included in the species list of the U.S. Fish and Wildlife Service Platte River Ecology Study summary report. Typical habitats are open woodlands, brushlands, and farmsteads.

Northern Mockingbird (*Mimus polyglottos*). An uncommon migrant and local summer resident in the eastern Platte Valley; at Mormon Island an accidental spring migrant, reported only in May. Typical habitats are fencerows and shrublands.

Gray Catbird (*Dumetella carolinensis*). A common migrant and summer resident in the Platte Valley; at Mormon Island a common to uncommon migrant and summer resident, and a probable breeder. Records extend from early May to mid-September. Typical habitats include riparian woodlands, shrublands, and fencerows.

Brown Thrasher (*Toxostoma rufum*). A common migrant and summer resident in the Platte Valley and at Mormon Island. Records extend from late April until late September. Typical habitats include fencerows, shrublands, and riparian woodlands.

American Robin (*Turdus migratorius*). A common to abundant migrant and summer resident in the Platte Valley, overwintering locally. At Mormon Island a common migrant and known breeder, uncommon in winter. Typical habitats include shrublands, riparian woodlands, wood lots, fencerows, farmsteads, and wetland meadows.

Swainson's Thrush (*Catharus ustulata*). A common migrant in the Platte Valley; at Mormon Island a rare spring migrant, reported only during May. Typical habitats are wood lots and riparian woodlands.

Gray-cheeked Thrush (*Catharus minimus*). A common migrant in the eastern Platte Valley, declining to the west. At Mormon Island an uncommon spring migrant, not reported during fall. Records are for May only. Typical habitats are wood lots.

Veery (*Catharus fuscescens*). An occasional to rare migrant in the Platte Valley; at Mormon Island a rare spring migrant. Records are for May only. Typical habitats are wood lots.

Eastern Bluebird (*Sialia sialis*). An uncommon migrant and summer resident in the eastern Platte Valley, declining to the west. At Mormon Island a rare migrant and summer visitor. Records are from mid-April until late September. Typical habitats are fencerows and riparian shrublands.

Golden-crowned Kinglet (*Regulus satrapa*). An uncommon to common migrant in the Platte Valley; not reported for Mormon Island by Lingle and Hays. Typical habitats are riparian shrublands and woodlands.

Ruby-crowned Kinglet (*Regulus calendula*). A common migrant and uncommon winter resident in the Platte Valley; at Mormon Island a rare migrant. Reports extend from September to April. Typical habitats include wood lots, idle lands, and riparian woodlands.

Northern Shrike (*Lanius excubitor*). An uncommon to locally common winter resident in the Platte Valley; at Mormon Island a rare winter resident. Typical habitats are riparian woodlands and open areas.

Loggerhead Shrike (*Lanius ludovicianus*). A common migrant and summer resident in the Platte Valley; overwintering rarely. At Mormon Island a rare migrant, reported only during April. Typical habitats are hay fields and other open areas with scattered perches.

European Starling (*Sturnus vulgaris*). An introduced common to abundant permanent resident in the Platte Valley, including Mormon Island. Typical habitats include wood lots, riparian woodlands, wetland meadows, croplands, and farmsteads.

Bell's Vireo (*Vireo bellii*). A common migrant and summer resident in the Platte Valley, declining to the west. At Mormon Island a rare migrant and summer resident, and a potential breeder. Reported during May and June. Typical habitats are shrublands and low woodlands.

Yellow-throated Vireo (*Vireo flavifrons*). An uncommon migrant and local summer resident in the eastern Platte Valley, decreasing to the west. Not reported for Mormon Island by Lingle and Hay. Typical habitats are mature riparian forests.

Red-eyed Vireo (*Vireo olivaceous*). A common migrant and locally common summer resident in the Platte Valley; at Mormon Island an occasional spring migrant. Reported only during April. Typical habitats are wood lots and mature riparian woodlands.

Warbling Vireo (*Vireo gilvus*). A common migrant and local summer resident in the Platte Valley, including Mormon Island. Records extend from early May until early September. Typical habitats include riparian woodlands, shrublands, wood lots, and fencerows.

Black-and-white Warbler (*Mniotilta varia*). A common migrant in the Platte Valley; at Mormon Island a rare spring migrant, reported only during May. Typical habitats are riparian woodlands.

Tennessee Warbler (*Vermivora peregrina*). A common migrant in the eastern Platte Valley, decreasing to the west. At Mormon Island an uncommon spring migrant, reported from late April to mid-May. Typical habitats include riparian woodlands and wood lots.

Orange-crowned Warbler (*Vermivora celata*). A common migrant in the Platte Valley; at Mormon Island an uncommon spring and common fall migrant. Records extend from mid-April to mid-May, and from late September to early October. Typical habitats are idle lands, fencerows, wood lots, and riparian woodlands.

Nashville Warbler (*Vermivora ruficapilla*). A common migrant in the Platte Valley, declining to the west. At Mormon Island a rare migrant, reported during mid-May and from early September to early October. Typical habitats include idle lands, shrublands, and wood lots.

Yellow Warbler (*Dendroica petechia*). A common migrant and summer resident in the Platte Valley, including Mormon Island. Records extend from early May until mid-June. Typical habitats include shrublands and riparian woodland edges.

Yellow-rumped Warbler (*Dendroica coronata*). A common migrant through the Platte Valley; at Mormon Island an uncommon migrant. Records extend from mid-April to mid-May, and from late September to early October. Typical habitats include idle lands, shrublands, and riparian woodlands.

Blackpoll Warbler (*Dendroica striata*). A common migrant in the eastern Platte Valley, declining to the west. At Mormon Island a rare spring migrant, reported only during May. Typical habitats are riparian woodlands and wood lots.

Ovenbird (*Seirus aurocapillus*). An uncommon to common migrant in the Platte Valley; at Mormon Island a rare spring migrant, recorded only during May. Typical habitats are wood lots and mature riparian woodlands.

Northern Waterthrush (*Seirus noveboracensis*). An uncommon migrant in the Platte Valley; at Mormon Island a rare spring migrant, recorded only during May. Typical habitats are riparian shrublands and damp woodland borders.

Common Yellowthroat (*Geothlypis trichas*). A common migrant and summer resident in the Platte Valley; at Mormon Island a common migrant and summer resident, potentially breeding. Records extend from late April to early September. Typical habitats are shrublands, riparian woodland thickets, wood lots, and marshy areas.

Yellow-breasted Chat (*Icteria virens*). A common migrant and local summer resident in the Platte Valley; not reported for Mormon Island by Lingle and Hall. Typical habitats include streamside thickets and riparian forest edges, as well as overgrazed pastures grown up to shrubs.

House Sparrow (*Passer domesticus*). An abundant permanent resident, introduced into Hall County in 1876. Typical habitats are farmsteads, idle lands, fencerows, and empty buildings.

Bobolink (*Dolichoynx oryzivorus*). A common migrant in the Platte Valley, and a locally common summer resident. At Mormon Island a common migrant and abundant breeding summer resident. Reports extend from early May to August. Typical habitats include wetland meadows, hay fields, alfalfa fields, and fencerows.

Eastern Meadowlark (*Sturnella magna*). A common migrant and local summer resident in the Platte Valley, especially to the east; at Mormon Island an abundant migrant and summer resident, often overwintering. Typical habitats include wetland meadows, hay fields, fencerows, alfalfa, and farmsteads.

Western Meadowlark (*Sturnella neglecta*). A common migrant and abundant summer resident in the Platte Valley; at Mormon Island a common migrant and a summer resident, potentially breeding. However, most breeding meadowlarks there are easterns. Frequently overwinters. Typical habitats are upland meadows, hay fields, fencerows, alfalfa and farmsteads.

Yellow-headed Blackbird (*Xanthocephalus xanthocephalus*). A common to abundant migrant and local summer resident in the Platte Valley; at Mormon Island a rare migrant, reported only during spring. Typical habitats include wetland meadows, farmsteads, fencerows, and riverine areas.

Red-winged Blackbird (*Agelaius pheoniceus*). A common to abundant migrant, winter visitor, and summer resident in the Platte Valley; at Mormon Island an abundant migrant and summer resident, often overwintering. Typical habitats include wetland meadows, croplands, shrublands, and riparian woodlands.

Orchard Oriole (*Icterus spurius*). A common migrant and summer resident in the Platte Valley; at Mormon Island a rare spring migrant and uncommon summer resident, potentially breeding. Records extend from mid-May to late June. Typical habitats are riparian woodlands.

Northern Oriole (*Icterus galbula*). A common migrant and summer resident in the Platte Valley and at Mormon Island, potentially breeding on the Island. Records extend from early May to early September. Typical habitats are wood lots and riparian woodlands.

Rusty Blackbird (*Euphagus carolinus*). A common to rare migrant in the Platte Valley, declining to the west, and a local winter resident. At Mormon Island a rare fall migrant, reported only during November. Typical habitats are riparian woodlands.

Brewer's Blackbird (*Euphagus cyanocephalus*). A variably common migrant in the Platte Valley, becoming more common to the west. At Mormon Island a rare spring migrant, reported only during April. Typical habitats are farmsteads.

Great-tailed Grackle (*Quiscalus mexicanus*). A rare migrant and a very rare and local summer resident in the Platte Valley; at Mormon Island a rare spring migrant and summer visitor. Records are from mid-May to mid-June. Typical habitats are shrublands, alfalfa fields, and riverine areas.

Common Grackle (*Quiscalus quiscula*). A common to abundant migrant and summer resident in the Platte Valley; at Mormon Island a common migrant and uncommon summer resident, potentially breeding. Records extend from mid-March to early October. Typical habitats include wood lots, hay fields, and farmsteads.

Brown-headed Cowbird (*Molothrus ater*). A common to abundant migrant and summer resident in the Platte Valley, including Mormon Island. Records extend from early April to late October. Typical habitats are wetland meadows, croplands, shrublands, and riparian woodlands; species parasitized on Mormon Island include red-winged blackbirds, bobolinks, eastern meadowlarks, dickcissels, and grasshopper sparrows.

Northern Cardinal (*Cardinalis cardinalis*). A common permanent resident in the eastern Platte Valley, declining to the west. At Mormon Island a rare permanent resident and potential breeder. Typical habitats include riparian woodlands, shrublands, and wood lots.

Black-headed Grosbeak (*Pheucticus melanocephalus*). A common migrant and summer resident in the western Platte Valley, declining to the east. Not listed for Mormon Island by Lingle and Hall, but included in the U.S. Fish and Wildlife Services' list of species identified in the Platte River Ecology Study. Typical habitats include riparian woodland edges and wood lots.

Rose-breasted Grosbeak (*Pheucticus ludovicanus*). A common migrant and summer resident in the eastern Platte Valley, declining to the west. At Mormon Island an uncommon spring migrant, not reported during fall, according to Lingle and Hay, but considered common in late spring by Short (*Nebraska Bird Review* 29:11). Typical habitats include riparian woodland and wood lots.

Blue Grosbeak (*Guiraca caerulea*). An uncommon migrant and local summer resident in the Platte Valley, not listed by Lingle and Hay for Mormon Island. Included in the U.S. Fish and Wildlife Services list of species observed in the Platte River Ecology Study. Typical habitats are weedy pastures, idle fields, and riparian forest edges.

Indigo Bunting (*Passerina cyanea*). An uncommon migrant and summer resident in the eastern Platte Valley, declining to the west. At Mormon Island a rare spring migrant, reported only during May. Typical habitats are riparian woodland edges and shrubland.

Lazuli Bunting (*Passerina amoena*). An uncommon migrant and summer resident in the western parts of the Platte Valley, declining to the east. Not mentioned by Lingle and Hay for Mormon Island, but among the species observed in the Fish and Wildlife Service's

Platte River Ecology Study. Typical habitats include shrublands and riparian woodland edges.

Dickcissel (*Spiza americana*). An abundant migrant and summer resident in the eastern Platte Valley, becoming uncommon to the west. At Mormon Island a common migrant and abundant breeding resident. Observed from mid-May through summer. Typical habitats are wetland meadows, hay fields, and alfalfa fields.

House Finch (*Carpodacus mexicanus*). A locally common permanent resident in the western Platte Valley, east to about North Platte. At Mormon Island a rare winter visitor seen only during December. Typical habitats include idle lands and sites around human habitations.

American Goldfinch (*Carduelis tristis*). A common permanent resident in the Platte Valley, at Mormon Island an abundant resident and probable breeder. Typical habitats include idle lands, shrublands, riparian woodlands, wood lots, wetland meadows, and croplands.

Rufous-sided Towhee (*Pipilo erythropthalmus*). A common migrant and local summer resident in the eastern Platte Valley, declining to the west. At Mormon Island an uncommon migrant, reported only during spring. Typical habitats are riparian woodland understories.

Lark Bunting (*Calamospiza melanocorys*). A common migrant and summer resident in the western Platte Valley, declining to the east. At Mormon Island a rare summer visitor, observed only during late May. Typical habitats are fencerows and open grasslands.

Savannah Sparrow (*Passerculus sandwichensis*). A common migrant in the Platte Valley, including Mormon Island. Observed in spring from late March to mid-May, and in fall from early September to late November. Typical habitats are fencerows, corn fields, alfalfa fields, and idle lands.

Grasshopper Sparrow (*Ammodramus savannarum*). A common migrant and abundant summer resident in Platte Valley, including Mormon Island. Records extend from early May to mid-October. Typical habitats are wetland meadows, hay fields, alfalfa fields, and fencerows. The most abundant breeding sparrow in grasslands.

Henslow's Sparrow (*Ammodramus henslowii*). An occasional migrant and possible local summer resident in the eastern Platte Valley, becoming rare to the west. At Mormon Island a rare fall migrant, recorded only during September. Typical habitats are fencerows, idle lands, and pastures.

Leconte's Sparrow (*Ammospiza leconteii*). An uncommon migrant in the eastern Platte Valley and rare or absent to the west; at Mormon Island a rare fall migrant. Reported only during October. Typical habitats are fencerows, wet meadows, and marshy edges.

Vesper Sparrow (*Pooecetes gramineus*). A common migrant and summer resident in the Platte Valley; at Mormon Island a common to abundant migrant. Reported from mid-April to early May, and from mid-September to early October. Typical habitats include fencerows, wetland meadows, and riparian woodlands.

Lark Sparrow (*Chondestes grammacus*). A common migrant and locally common summer resident in the Platte Valley; at Mormon Island an uncommon spring migrant and a rare summer visitor. Records are for May only. Typical habitats are fencerows, weedy fields with small trees, and woody edges.

Cassin's Sparrow (*Aimophila cassinii*). A rare or irregular migrant in the western Platte Valley; absent to the east. Not reported for Mormon Island by Lingle and Hay, but included in the U.S. Fish and Wildlife Service's Platte River Ecology Study summary report.

Dark-eyed Junco (*Junco hyemalis*). A common migrant and winter visitor in the Platte Valley; at Mormon Island an uncommon to common winter visitor and migrant. Records are from late September to late April. Typical habitats include idle lands, wood lots, and shrublands.

American Tree Sparrow (*Spizella arborea*). A common migrant and winter visitor in the Platte Valley; at Mormon Island an uncommon to abundant migrant and winter visitor. Records extend from late October to late March. Typical habitats include idle lands, fencerows, riparian woodlands, shrublands, and wood lots.

Chipping Sparrow (*Spizella passerina*). A common migrant and summer resident in the Platte Valley; at Mormon Island a variable common migrant. Records are from late April to mid-May, and for September. Typical habitats include fencerows, wood lots, croplands, and riparian woodland.

Clay-colored Sparrow (*Spizella pallida*). A common migrant in the Platte Valley; a single nesting record from Hall County. At Mormon

Island a common spring migrant and rare fall migrant. Records are from late April to mid-May and for October. Typical habitats include idle lands and fencerows.

Field Sparrow (*Spizella pusilla*). A common migrant and locally common summer resident in the Platte Valley; at Mormon Island an uncommon migrant and summer resident, potentially breeding. Records are from late April to early October. Typical habitats include shrublands, riparian woodland edges, and overgrown pastures.

Brewer's Sparrow (*Spizella breweri*). A common migrant in the western parts of the Platte Valley; absent to the east. Not listed for Mormon Island by Lingle and Hay, but included in the U.S. Fish and Wildlife Service's Platte River Ecology Study summary report. Typical habitats are brushlands and scrubby grasslands.

Harris Sparrow (*Zonotrichia querula*). A common to abundant migrant and winter visitor in the Platte Valley; at Mormon Island uncommon in winter and spring, but common in fall. Records are from mid-October to mid-May.

White-crowned Sparrow (*Zonotrichia leucophrys*). A common migrant and winter visitor in the Platte Valley; at Mormon Island an uncommon spring and fall migrant, recorded from late April to late May, and from late September to early November. Typical habitats include idle lands, fencerows, wood lots, and shrublands.

White-throated Sparrow (*Zonotrichia albicollis*). A common migrant and winter visitor in the Platte Valley; at Mormon Island a rare migrant in spring and fall. Recorded only for April, October and November. Typical habitats are idle lands, wood lots, and riparian woodlands.

Lincoln's Sparrow (*Melospiza lincolnii*). A common migrant in the Platte Valley; at Mormon Island an uncommon fall migrant, not reported during spring. Records are from early September to mid-October. Typical habitats are idle lands and fencerows.

Swamp Sparrow (*Melospiza georgiana*). An uncommon migrant in the eastern Platte Valley, becoming rare to the west, and a rare and local summer resident. At Mormon Island a rare fall migrant, reported only during October. Typical habitats are idle lands, wet meadows, and marshy areas.

Song Sparrow (*Melospiza melodia*). A common migrant and winter visitor, and a local summer resident in the Platte Valley (including Hall County); at Mormon Island a common migrant, a rare summer visitor, and an uncommon winter resident. Typical habitats include shrublands, fencerows, idle lands, and riparian woodlands.

Chestnut-collared Longspur (*Calcarius ornatus*). A migrant and winter resident in the Platte Valley, especially to the west. Not listed for Mormon Island by Lingle and Hay, but included in the U.S. Fish and Wildlife Service's list of species identified in the Platte River Ecology Study. Typical habitats are open grassy fields.

The Platte River Ecology Study of the U.S. Fish and Wildlife Service (1981) reported that their study area (316 townships, or 1,896 square miles) of the central Platte Valley supported an estimated 142 species of breeding birds, or about 70 percent of all the species of birds currently known to breed within the state's boundaries. Of these, 34 species accounted for about 95 percent of the total estimated population, with western meadowlarks, common grackles, and grasshopper sparrows collectively constituting about 26 percent. House sparrows, mourning doves, brown-headed cowbirds, and red-winged blackbirds added another 24 percent. Species that individually constituted between 4.1 and 1.0 percent of the population were (in descending frequency) the American robin, lark bunting, western kingbird, chimney swift, horned lark, starling, dickcissel, eastern kingbird, barn swallow, northern oriole, lark sparrow, common flicker, house wren, bobolink, upland sandpiper, and red-headed woodpecker. Species individually constituting from 0.9 to 0.2 percent included the brown thrasher, orchard oriole, cliff swallow, killdeer, yellow warbler, American goldflinch, ring-necked pheasant, common yellowthroat, common nighthawk, bobwhite, and blue jay.

This study concluded that shelterbelts supported the highest densities of breeding birds (1,361 pairs per 100 acres) and were used by 31 breeding species. River channel islands had an estimated mean density of 212.4 breeding pairs per 100 acres, and 35 probable breeding species. Lowland riverine forests supported a total of 50 breeding species, with a mean breeding density of 202.4 pairs per 100 acres. Lowland native prairie supported 27 estimated breeding species, and a mean density of 47.4 pairs per 100 acres. Upland native prairie supported an estimated 31 breeding species, and a mean density of 39.3 pairs per 100 acres. Alfalfa fields supported 13 nesting species, with a mean density of 41.3 pairs per 100 acres. Corn fields supported three breeding species at a mean density of 15.3 pairs per 100 acres, and wheat fields supported 13 breeding species, at a mean density of 13.4 pairs per 100 acres.

Checklist of Mammals of the Platte Valley[1]

Opossum (*Didelphis marsupialis*).* Widespread in S. Platte and Platte River valleys; habitats are riparian woods and sometimes also urban areas.

Masked Shrew (*Sorex cinereus*).* Widespread over entire valley, occurring in various moist habitats.

Short-tailed Shrew (*Blarina brevicauda*).* In middle and lower Platte Valley west to Lincoln County; diverse habitats, usually in wet areas of woods or weedy fields.

1. Excluding bats and extirpated species, mainly after Knox Jones (1964). Species reported from Mormon Island are indicated by asterisks (Lingle and Boner, 1981).

Least Shrew (*Cryptotis parva*). Rare throughout entire valley, in open grasslands, brush, and dry fields.

Eastern Mole (*Scalopus aquaticus*).* Throughout entire valley, especially in habitats with moist but well-drained and fairly loose soils.

Black-tailed Jackrabbit (*Lepus californicus*).* Throughout entire valley, especially pastures, haylands, and native grasslands.

Eastern Cottontail (*Sylvilagus floridanus*).* Throughout entire valley, typically in brushy habitats or woodland edges.

Desert Cottontail (*Sylvilagus auduboni*). In North Platte valley east to Garden County, in high plains grassland, with scattered brush.

Eastern Fox Squirrel (*Sciurus niger*).* Throughout entire valley, in woods, wood lots, farmsteads, and urban areas.

Woodchuck (*Marmota monax*). Local in lower Platte valley west to Saunders County; typical habitats are woodland edges, often near water.

Black-tailed Prairie Dog (*Cynomys ludovicianus*).* Upper and middle Platte valley, east to Hall County and at least formerly to Platte County; in dry grasslands.

Thirteen-lined Ground Squirrel (*Spermophilus tridecemlineatus*).* Throughout entire valley, in open grasslands, and sometimes also woodland borders.

Spotted Ground Squirrel (*Spermophilus spilosoma*). N. and S. Platte valleys east to Dawson County, in dry grasslands.

Franklin Ground Squirrel (*Spermophilus franklini*).* In N. Platte and Platte valleys eastward from Morrill County, in grass-woodland edge areas.

Eastern Chipmunk (*Tamias striatus*). In lower Platte valley west to Dodge County; usually in woodland edges.

Southern Flying Squirrel (*Glaucomys volans*). In lower Platte valley west to Dodge County; usually in heavy timber near water.

Plains Pocket Gopher (*Geomys bursarius*).* Throughout entire valley; in open grasslands, pastures, and hay fields.

Olive-backed Pocket Mouse (*Perognathus fasciatus*). In N. Platte valley east to Garden County; associated with dry grasslands.

Plains Pocket Mouse (*Perognathus flavescens*). Almost throughout entire valley except easternmost areas; associated with dry grasslands.

Silky Pocket Mouse (*Perognathus flavus*). In N. Platte valley east to Morrill County; rare in dry grasslands.

Hispid Pocket Mouse (*Perognathus hispidus*).* Throughout entire valley; associated with dry grasslands.

Ord Kangaroo Rat (*Dipodomys ordi*). The upper and middle Platte valleys east to Platte County; in sandy grasslands.

Beaver (*Castor canadensis*).* Throughout entire valley; associated with rivers and streams.

Plains Harvest Mouse (*Reithrodontomys montanus*).* Throughout entire valley; in upland grasslands.

Western Harvest Mouse (*Reithrodontomys megalotis*).* Throughout entire valley; in grasslands and weedy areas, usually near water.

Deer Mouse (*Peromyscus maniculatus*).* Throughout entire valley; usually in grasslands and other nonwooded areas.

White-footed Mouse (*Peromyscus leucopus*).* Middle and lower Platte valley west to Lincoln County; primarily in wooded areas.

Northern Grasshopper Mouse (*Onychomys leucogaster*). Most or all of upper and middle Platte valley, but becomes rare east of Lincoln County; typically in open grassland habitats.

Eastern Woodrat (*Neotoma floridana*). Upper Platte valley east to Lincoln County; usually in rocky timbered areas.

Southern Bog Lemming (*Synaptomys cooperi*). Highly local in lower Platte valley west to Butler County; usually in heavy grassy habitats and moist sites.

Muskrat (*Ondatra zibethicus*). Throughout entire valley; associated with water.

Meadow Vole (*Microtus pennsylvanicus*).* Throughout entire valley; mainly in moist and grassy habitats.

Prairie Vole (*Microtus ochrogaster*). Throughout entire valley; in upland grassy or thickety areas.

Meadow Jumping Mouse (*Zapus hudsonius*).* Throughout entire valley; usually in open grassy habitats near water, sometimes in woodland edges.

Norway Rat (*Rattus norvegicus*). Introduced throughout entire valley; in farmsteads, urban areas and other human-associated habitats.

House Mouse (*Mus musculus*).* Introduced throughout entire valley; in farmsteads, urban areas, and other human-associated habitats.

Porcupine (*Erethizon dorsatum*). Upper N. Platte valley, associated with woodlands.

Coyote (*Canis latrans*).* Entire valley, generally distributed but typical of open country.

Red Fox (*Vulpes vulpes*). Entire valley, becoming rarer to the west; in mixed forests and grassland areas.

Swift Fox (*Vulpes velox*). Formerly throughout entire upper valley; now probably extirpated from Nebraska except in the northwestern Panhandle.

Gray Fox (*Urocyon cinereoargenteus*). Entire valley, but becoming rarer to the west; in wooded areas and brushlands, but avoiding dense, mature forests.

Raccoon (*Procyon lotor*).* Throughout entire valley; typically in riparian woodlands.

Least Weasel (*Mustela rixosa*).* Middle and lower Platte valley; generally distributed but rare.

Long-tailed Weasel (*Mustela frenata*). Entire valley; generally distributed but especially in brushy areas or woods near water, with some open areas.

Mink (*Mustela vison*). Throughout entire valley, near permanent water.

Badger (*Taxidea taxus*).* Throughout entire valley; generally in open grasslands.

Striped Skunk (*Mephitis mephitis*).* Throughout entire valley; generally distributed but preferring brushy or woodland areas.

Eastern Spotted Skunk (*Spilogale putorius*). Local throughout entire valley; usually in open grasslands or brushy areas.

Bobcat (*Lynx rufus*). Local in the upper Platte area, especially to the west; usually in heavy wooded cover with rocky cliffs nearby.

Mule Deer (*Odocoileus hemionus*).* Throughout entire valley, becoming more common to the west; favoring grassland and brushy areas of irregular topography.

White-tailed Deer (*Odocoileus virginianus*).* Throughout entire valley, becoming more common to the east; favoring timbered bottom lands.

Pronghorn (*Antilocapra americana*). In upper Platte Valley east to about Keith County; favoring open grassy plains.

Checklist of Amphibians and Reptiles of Hall County[1]

Reported from Mormon Island Preserve

Striped Chorus Frog (*Pseudacris triseriata*). Common in ditches and other areas of standing water.

Woodhouse's Toad (*Bufo woodhousii*). Common in sandy areas, such as along stream banks.

Plains Leopard Frog (*Rana blairi*). Common in areas of permanent standing water; the most abundant amphibian on Mormon Island.

1. Adapted from Jones, Ballinger, and Nietfeldt (1981).

Snapping Turtle (*Chelydra serpentina*). Common on large ponds and in the Platte River.

Spiny Softshell (*Tionyx spiniferus*). Common in the Platte River.

Painted Turtle (*Chrysemys picta*). Common in all water areas.

Prairie Skink (*Eumeces septentrionalis*). Common in streamside woodland having logs, flat rocks, or other debris.

Six-lined Racerunner (*Cnemidophorus sexlineatus*). Common in sandy areas.

Plains Garter Snake (*Thamnophis radix*). Common in wet meadows and streamside woodlands, sometimes in prairie areas.

Common Garter Snake (*Thamnophis sirtalis*). Most common near pools and along the Platte River.

Not Seen on Mormon Island but Reported from Hall County

Great Plains Toad (*Bufo cognatus*). Found in large temporary ponds in drier habitats.

Bullfrog (*Bufo catesbeiana*). Typical of extensive marshy areas.

Cricket Frog (*Acris crepitans*). Found along the Platte River just east of Mormon Island.

Gray Treefrog (*Hyla chrysocelis*). In permanent water areas among woodlands, west to Hall County.

Ornate Box Turtle (*Terrapene ornata*). In open, sandy areas.

Racer (*Coluber constrictor*). Primarily in grassland or open, brushy woods.

Bullsnake (*Pituophis catenifer*). Primarily in grassland habitats.

Lined Snake (*Tropidoclonion lineatum*). Nocturnal; mainly found under logs or underground.

Checklist of Platte River Fishes of Nebraska[1]

Shovelnose Sturgeon (*Scaphirhynchus platorynchus*). Lower Platte west to Columbus.

Paddlefish (*Polyodon spathula*). Lower Platte west to Fremont.

Longnose Gar (*Lepisosteus osseus*). Platte west to Kearney.

Shortnose Gar (*Lepisosteus platostomus*).* Lower Platte west to Columbus.

1. Mainly derived from Morris, Morris, and Witt (1972), and John Lynch (personal comment). Asterisks indicate species reported from Mormon island by Cochran and Jenson (1981); daggers indicate species reported from Shoemaker Island by Bliss and Schainost (1973).

Gizzard Shad (*Dorosoma cepedianum*).*† Platte system west to above Lake McConaughy.
Goldeye (*Hiodon alosoides*).*† Platte west to Kearney.
Rainbow Trout (*Salmo gairdneri*). Introduced into N. Platte R.
Brown Trout (*Salmo trutta*). Introduced into N. Platte R.
Stoneroller (*Campstoma anomalum*). Entire Platte system.
Brassy Minnow (*Hybognathus hankinsoni*).*† Entire Platte system.
Western Silvery Minnow (*Hybognathus argyritis*).*† Platte west to North Platte (city).
Plains Minnow (*Hybognathus placitus*).† Entire Platte system.
Speckled Chub (*Hybopsis aestivalus*).* Platte west to Kearney.
Sturgeon Chub (*Hybopsis gellda*). Platte west to North Platte.
Flathead Chub (*Hybopsis gracilis*).† Platte west to North Platte (city).
Silver Chub (*Hybopsis storeriana*). Lower Platte west to Grand Island.
Golden Shiner (*Notemigonus crysoleucas*). Lower Platte west to Grand Island; widely introduced.
Emerald Shiner (*Notropis atherinoides*).*† Lower Platte west to Grand Island; introduced at Lake McConaughy.
River Shiner (*Notropis blennis*).* Platte and N. Platte R. west to Keystone.
Common Shiner (*Notropis cornutus*). Scottsbluff area, Lodgepole Creek, South Platte, and extreme lower Platte.
Bigmouth Shiner (*Notropis dorsalis*).† Entire Platte system.
Red Shiner (*Notropis lutrensis*).*† Entire Platte system.
Spottail Shiner (*Notropis hudsonianus*). Introduced at Lake McConaughy.
Sand Shiner (*Notropis stramineus*).*† Entire Platte system.
Suckermouth Shiner (*Pnenacobius mirabilis*). Spotty in Platte system.
Fathead Minnow (*Pimaphales promelas*).*Entire Platte system.
Longnose Dace (*Rhinichthys cataractae*). North Platte R. to North Platte.
Northern Redbelly Dace (*Phaxinos eos*). A few streams debauching into N. Platte R. between Keystone and Sutherland.
Creek Chub (*Semotilus atromaculatus*).* Entire Platte system.
River Carpsucker (*Carpoides carpio*).*† Entire Platte system.

Highfin Carpsucker (*Carpoides velifer*). Lower Platte to Fremont.

Quillback (*Carpoides cyprinus*).*† Entire Platte system.

Longnose Sucker (*Catostomus catostomus*). Platte R. east to Keystone, and Lodgepole Creek.

White Sucker (*Catostomus commersoni*).* Entire Platte system.

Bigmouth Buffalo (*Ictiobus cyprinellus*). Platte west to Kearney.

Northern Redhorse (*Moxostoma macrolepidotum*).† Entire Platte system.

Carp (*Cyprinus carpio*).† Introduced into entire Platte system.

Black Bullhead (*Ictalurus melas*).*† Entire Platte System.

Yellow Bullhead (*Ictalurus natalis*). Platte west to North Platte (city).

Channel Catfish (*Ictalurus punctatus*).*† Entire Platte system.

Flathead Catfish (*Pylodictis olivaris*). Lower Platte; spotty on upper Platte and North Platte.

Stonecat (*Noturus flavus*). Entire Platte system.

Tadpole Madtom (*Noturus gyrinus*). Lower Platte west to Columbus.

Plains Topminnow (*Fundulus sciadicus*). Entire Platte system.

Plains Killifish (*Fundulus zebrinus*).* Entire Platte system.

Brook Stickleback (*Culaea inconstans*). Spotty in Platte system.

White Bass (*Morone chrysops*).* Introduced through entire system.

Striped Bass (*Morone saxatilis*). Introduced into Lake McConaughy.

White Perch (*Morone americana*).† Lower Platte west rarely to Hall County.

Bluegill (*Lepomis macrochirus*).* Entire Platte system.

Rock Bass (*Ambloplites rupestris*).† Introduced into middle Platte.

Green Sunfish (*Lepomis cyanellus*).*† Entire Platte system.

Orangespotted Sunfish (*Lepomis humilis*).* Platte system west to Oshkosh and Big Springs.

White Crappie (*Pomoxis annularis*).* Entire Platte system.

Black Crappie (*Pomoxis nigromaculatus*).*† Entire Platte system.

Smallmouth Bass (*Micropterus dolomieui*).*† Introduced through most of Platte system.

Largemouth Bass (*Micropterus salmoides*).*† Introduced through most of Platte system.

Johnny Darter (*Etheostoma nigrum*). Scattered through the Platte system.

Iowa Darter (*Etheostoma exile*). Scattered through the Platte system.

Yellow Perch (*Perca flavescens*). Widely introduced in upper Platte R.
Sauger (*Stizostedion canadense*). Lower Platte west to Loup R.
Walleye (*Stizostedion vitreum*). Entire Platte system.
Freshwater Drum (*Aplodinotus grunniens*).* Platte and N. Platte R. west to Oshkosh.

Checklist of Common Plants of the Platte River Flood Plain[1]

Trees

American Elm (*Ulmus americana*). In rich bottomlands.
Black Locust (*Robinia pseudoacacia*). On moist, well-drained soils.
Black Walnut (*Juglans nigra*). On moist, rich soils.
Box Elder (*Acer negundo*). On moist, rich soils.
Catalpa (*Catalpa speciosa*). On moist, rich soils.
Chinese Elm (*Ulmus parvifolia*). An introduced species.

1. Adapted from U.S. Fish and Wildlife Service (1981).

Choke Cherry (*Prunus virginiana*). On moist, rich soils.

Cottonwood (*Populus deltoides*). A widespread species, especially near water.

Diamond Willow (*Salix rigida*). On damp soils.

Green Ash (*Fraxinus pennsylvanica*). On moist, well-drained soils.

Hackberry (*Celtis occidentalis*). On moist, well-drained soils.

Osage Orange (*Maclura pomifera*). A widely planted species.

Peach-leaf Willow (*Salix amygdaloides*). On wet soils.

Red Cedar (*Juniperus virginiana*). Occasionally in streamside forests.

Red Mulberry (*Morus rubra*). On rich bottomland soils.

Red-osier Dogwood (*Cornus stolonifera*). On wet soils.

Russian Olive (*Elaeagnus angustifolia*). A planted species.

Salt Cedar (*Tamarix ramosissima*). A self-introduced species; found on river bars.

Sandbar Willow (*Salix exigua*). On sandbars and islands.

Siberian Elm (*Ulmus pumila*). An introduced species.

Silver Maple (*Acer saccharinum*). On bottom land soils.

Slippery Elm (*Ulmus rubra*). In rich bottom lands.

Shrubs and Vines

Buffaloberry (*Shepherdia argentea*). On streambanks.

Buckbrush (*Symphoricarpos orbiculatus*). In woods and old fields.

Climbing Bittersweet (*Celastrus scandens*). In woods.

Prickly Ash (*Zanthoxylum americanum*). In damp woods.

Rough-leaf Dogwood (*Cornus drummondii*). On wet soils.

Skunkbush (*Rhus aromatica*). On sandy soils.

Smooth Sumac (*Rhus glabra*). In openings of woods.

Snowberry (*Symphoricarpos orbiculatus*). In prairies.

Virginia Creeper (*Parthenocissus vitacea*). In woods.

Wild Grape (*Vitis riparia*). On streambanks.

Wild Plum (*Prunus americana*). In thickets.

Wild Rose (*Rosa woodsii*). In prairies.

Woodbine (*Parthenocissus vitacea*). In woods.

Grasses and Sedges

American Bulrush (*Scirpus americanus*). In wet areas.
Barnyard Grass (*Echinochloa crus-galli*). In disturbed areas.
Big Bluestem (*Andropogon gerardi*). In tall prairies.
Blue Grama (*Bouteloua gracilis*). In prairies.
Bluegrass (*Poa pratensis*). In pastures, introduced.
Buffalo Grass (*Buchloe dactyloides*). In dry grasslands.
Canada Wild Rye (*Elymus canadensis*). In wet prairies.
Cattail (*Typha latifolia*). In wet areas.
Cheatgrass (*Bromus tectorum*). In disturbed areas.
Corn (*Zea mays*). A planted crop species.
Foxtail Barley (*Hordeum jubatum*). An introduced weed.
Hairy Grama (*Bouteloua hirsuta*). In dry grasslands.
Hardstem Bulrush (*Scirpus acutus*). In wet areas.
Indian Grass (*Sorghastrum avenaceum*). In tall prairies.
Japanese Brome (*Bromus japonicus*). An introduced species; various habitats.
Junegrass (*Koeleria cristata*). In prairies.
Little Bluestem (*Andropogon scoparius*). In prairies.
Lovegrass (*Eragrostis pectinacea*). On disturbed areas.
Milo (*Sorghum sp.*). A planted crop species.
Needle and Thread (*Stipa comata*). In prairies.
Needle-leaf Sedge (*Carex eleocharis*). In meadows.
Nutsedge (*Cyperus sp.*). In meadows.
Plains Muhly (*Muhlenbergia cuspidata*). In prairies.
Prairie Cordgrass (*Spartina pectinata*). In wet areas.
Redtop (*Agrostis stolonifera*). In moist areas.
Reed Canary Grass (*Phalaris arundinacea*). In wet areas.
Saltgrass (*Distichlis stricta*). In alkaline areas.
Sand Dropseed (*Sporobolus cryptandrus*). In sandy areas.
Sedges (*Carex sp.*). In meadows.
Side Oats Grama (*Bouteloua curtipendula*). In prairies.
Six-week Fescue (*Festuca octoflora*). In disturbed areas.
Small Panic Grass (*Panicum oligosanthes*). In sandy, moist woods.
Smooth Brome (*Bromus inermis*). An introduced species.
Spikerush (*Eleocharis palustria*). In wet areas.
Switchgrass (*Panicum virgatum*). In prairies.

Water Sedge (*Carex aquatilis*). In wet areas.
Western Wheatgrass (*Agropyron smithii*). In prairies.
Wheatgrass (*Agropyron sp.*). In prairies.

Broad-leaved Herbs (Forbs)

Alfalfa (*Medicago sativa*). A planted crop species.
American Bugleweed (*Lycopus americanus*). In wet ground near streams.
American Germander (*Teucrium canadense*). In woods.
Aster (*Aster sp.*). In prairies and wetland shrub habitats.
Black-eyed Susan (*Rudbeckia hirta*). In prairies, disturbed areas.
Blazing Star (*Liatris punctata*). In prairies.
Bundleflower (*Desmanthus illinoensis*). In prairies, especially sandy areas.
Canada Goldenrod (*Solidago canadensis*). In prairies, open woods.
Canada Sanicle (*Sanicula canadensis*). In woods.
Cocklebur (*Xanthium strumarium*). In disturbed areas.
Common Mullein (*Verbascum thapsus*). In disturbed areas.
Common Ragweed (*Ambrosia artemisiifolia*). In disturbed areas.
Cone Flower (*Ratibida columnaris*). In prairies.
Dandelion (*Taraxacum officinale*). In meadows, disturbed areas.
Fringed Loosestrife (*Lysimachia ciliata*). In prairies.
Fog Fruit (*Phyla lanceolata*). On very moist ground.
Ground Cherry (*Physalis virginiana*). In disturbed areas.
Hoary Vervain (*Verbena stricta*). In disturbed areas.
Hopseed (*Medicago lupulina*). In disturbed areas.
Indigo Bush (*Amorpha fruticosa*). In prairies.
Ironweed (*Vernonia fasciculata*). In prairies, pastures.
Lady's Thumb Smartweed (*Polygonum persicaria*). In disturbed areas.
Lead Plant (*Amorpha canescens*). In prairies.
Loosestrife (*Lythrum dacotanum*). In prairies.
Pale Smartweed (*Polygonum lapathifolium*). In wet areas.
Plains Sunflower (*Helianthus petiolaris*). In prairies.
Poison Ivy (*Rhus radicans*). In open woods.
Prairie Dogbane (*Apocynum sibiricum*). In waste places.
Red Clover (*Trifolium pratense*). A planted crop species.

Scarlet Gaura (*Gaura coccinea*). In prairies, disturbed areas.
Skeleton Weed (*Lygodesmia juncea*). In prairies, disturbed areas.
Snow-on-the-Mountain (*Euphorbia marginata*). On prairies, disturbed areas.
Spikenard (*Aralia racemosa*). In woods.
Spotted Spurge (*Euphorbia maculata*). In disturbed areas.
White Sweet Clover (*Melilotus alba*). A planted crop species.
Wild Alfalfa (*Psoralea tenuiflora*). In prairies.

A male wood duck in a cottonwood tree. Photo by author

Bibliographic Notes

1. The Past

The Land. An account of the faunal changes associated with the end of the Mesozoic era can be found in D. A. Russell, 1982, "The mass extinction of the late Mesozoic," *Scientific American* 246(1):58–65. A useful reference to fossil life of the high plains is "Fossils of Wyoming," by M. W. Hager, 1971, *Geological Survey of Wyoming Bulletin* 54. A reference relating to Nebraska is A. L. Lugn, 1934, "Outline of Pleistocene geology of Nebraska: Part 1. Geology and mammalian fauna of the Pleistocene of Nebraska," *University of Nebraska*

State Museum Bulletin 1, no. 41, pt. 1. A discussion of the unique fossil fauna of western Nebraska may be found in the National Park Service, 1980, "Agate fossil beds national monument," Handbook 107, U.S. National Park, Washington, D.C. M. R. Voorhies, 1981, described the Niobrara Valley Poison Ivy quarry in "Ancient ashfall creates a Pompeii of prehistoric animals," *National Geographic* 159(1):66–75. Other useful geological references include the *Guidebook to the late Pliocene and early Pleistocene of Nebraska,* by T. M. Stout, *et al.,* 1971, Lincoln: University of Nebraska and Survey Division, Nebraska Geological Survey; and "Revision of the classification of the Pleistocene deposits of Nebraska," by E. C. Reed and V. H. Dreeszen, 1965, *Nebraska Geological Survey Bulletin* no. 23.

A more general but highly valuable reference is that of W. P. Webb, *The Great Plains,* Boston: Ginn and Co. (reprinted in 1981 by the University of Nebraska Press, Lincoln). A discussion of both extinct and modern species of bison may be found in J. S. McDonald, 1981, *North American Bison: Their Classification and Evolution,* Berkeley: University of California Press. An excellent review of Pleistocene mammals is *Pleistocene Mammals of North America,* by B. Kurten and E. Anderson, 1980, New York: Columbia University Press. Pleistocene mammals of Nebraska are discussed in "The rescue of Ice Age mammals," by M. Voorhies, R. Corner, and L. Tanner, 1982, *Nebraskaland* 60(5):6–14.

The First People. A standard reference on the early human occupation of the Great Plains is W. R. Wedel, 1961, *Prehistoric Man on the Great Plains,* Norman: University of Oklahoma Press. An excellent general reference on the Pawnees is G. E. Hyde, 1971, *The Pawnee Indians,* Norman: University of Oklahoma Press. Two other more general references on Plains Indians are C. Wissler, 1934, *North American Indians of the Plains,* American Museum of Natural History Handbooks, no. 1; and G. E. Hyde, 1959, *Indians of the High Plains: from the Prehistoric Period to the Coming of the Europeans,* Norman: University of Oklahoma Press. The archaeology of the Pawnees is summarized by W. R. Wedel, 1936, "An introduction to Pawnee archeology," *Bulletin* 112, U.S. Bureau of American Ethnology. Archeological sites in Nebraska are discussed in M. F. Kivett, 1952, "Woodland sites in Nebraska," *Nebraska Historical Society*

Publications in Anthropology 1:1–102; and R. T. Grange, Jr., 1980, described the late bison hunter sites in "Archeological investigations in the Red Willow Basin," *Nebraska Historical Society Publications in Anthropology* 9:1–237. Ceramics of Nebraska Indians are described by R. T. Grange, Jr., 1968, in "Pawnee and Lower Loup pottery," *Nebraska Historical Society Publications in Anthropology* 3:1–236. Pawnee folklore is covered by G. B. Grinnell's *Pawnee Hero Stories and Folk-tales,* New York: Forest and Stream Pub. (reprinted in 1961 by University of Nebraska Press, Lincoln); and is summarized in G. A. Dorsey, 1906, "The Pawnee Mythology," Carnegie Institution, Publication 59. G. Weltfish, 1977, *The Lost Universe: Pawnee Life and Culture,* Lincoln: University of Nebraska Press, summarizes Pawnee culture. Their music is detailed in F. Densmore, 1929, "Pawnee music," *Bulletin* 93, Bureau of American Ethnology. The Hako ceremony is monographed in A. C. Fletcher, 1900–1901, "The Hako: A Pawnee ceremony," *Annual Report, U.S. Bureau of American Ethnology,* Washington. The Pawnee songs heading "The Past," "The Land," and the Appendix are from the Densmore reference; the Pawnee creation story is from Dorsey.

The River. The Pawnee songs heading and terminating the section are from the Densmore reference above. The account of the early history of the Platte River is based on E. O. Staley and W. J. Wayne, 1972, "Epeirogenic and climatic controls of early Pleistocene fluvial sediment dispersal in Nebraska," *Geological Society of America Bulletin* 83:3675–90.

Although not related specifically to the Platte, a general vegetational history of the area is H. E. Wright, 1970, "Vegetational history of the central plains," pp. 157–72, in *Pleistocene and Recent Environments of the Central Great Plains,* W. Dort and J. K. Jones, eds., Lawrence: University of Kansas Press.

The Next People. The standard reference on the use of the Platte River by immigrants is J. J. Mattes, 1969, *The Great Platte River Road,* Nebraska Historical Society Publications, vol. 25. Two valuable references on the Grand Island area are A. F. Beuchler, (ed.), 1920, *History of Hall County,* Lincoln: Western Publishing and Engraving Co.; and W. Stolley, 1946, *History of the First Settlement of Hall*

County, Nebraska, Lincoln: Nebraska State Historical Society, vol. 27 (special issue). Memories of a homesteading family in the Mormon Island area are found in C. Converse, 1963, *Island in the Prairies,* privately published. Finally, J. C. Olson, 1966, *History of Nebraska,* 2d ed., Lincoln: University of Nebraska Press, is also a valuable source.

The Island. The Pawnee creation story heading the section is from the reference by G. Dorsey, cited earlier. Plant communities in the Platte Valley and elsewhere in Nebraska are discussed in various papers and books by J. E. Weaver, including *Native Vegetation of Nebraska,* Lincoln: University of Nebraska Press, 1965. Recent observations of succession on river bars and islands are provided in the U.S. Fish and Wildlife Service's "The Platte River ecology study" (see complete reference in Appendix notes). Of related interest is P. J. Rand, 1973, "The woody phreatophyte communities of the Republican River Valley in Nebraska," unpublished report, Botany Department, University of Nebraska–Lincoln; and J. Nickel, 1978, "Origin and plant succession of Platte River islands," unpublished report, Kearney State College Department of Botany. A similar account is provided by P. J. Currier, 1981, "The floodplain vegetation of the Platte River: phytosociology, forest development and seedling establishment," Ph.D. dissertation, Iowa State University, Ames.

2. *The Present*

The Sandbar. The ecology of the cranes in Nebraska is discussed extensively in the U.S. Fish and Wildlife Service's "The Platte River ecology study" (see complete reference in Appendix notes), and in C. R. Frith, 1974, "The ecology of the Platte River as related to sandhill cranes and other waterfowl in south central Nebraska," M.S. thesis, Kearney State College. Also relevant is the article "Sandhill cranes: Wings over the Platte," by J. Farrar and K. Bouc, originally published in *Nebraskaland,* February 1980, pp. 19–34. The breeding behavior of the least tern is admirably summarized by E. Schulenberg, J. Schulenberg, and M. Schulenberg, 1980, "Distribution and ecological study of the least tern in Kansas," unpublished nongame wildlife project report to the Kansas Fish and Game Commission. Two useful references on the six-lined racerunner are H. S. Fitch, 1958, "Natural history of the six-lined racerunner (*Cnemidophorus sexlineatus*)," *University of Kansas Museum of Natural History* 11:11–62; and D. F. Hardy, 1962, "Ecology and behavior of the six-lined racerunner, *Cnemidophorus sexlineatus*," *University of Kansas Science Bulletin* 43:3–73.

The Shoreline Forest. The winter ecology of eagles on the Platte River is discussed in W. E. Vian, 1971, "The wintering bald eagle (*Haliaeetus leucocophalus*) on the Platte River in south-central Nebraska," M.S. thesis, Kearney State College. A useful reference on wood duck biology is "The wood duck in Massachusetts," by D. Grice and J. P. Rogers, 1965, Massachusetts Division of Fisheries and Game, Final Report W-19-R.

The Wet Meadow. Among the many studies of red-winged blackbird reproductive biology is R. E. Nero, 1956, "A behavior study of the red-winged blackbird," *Wilson Bulletin* 68:5–37, 129–50. As in the other essays, my discussions of individual species are essentially fictionalized accounts based on personal observations and especially on various published descriptions of breeding biology. The major reference used for the long-billed marsh wren account was J. Verner, 1965, "Breeding biology of the long-billed marsh wren," *Condor* 67: 6–30.

The Wood Lot. The account of the red-tailed hawk (including the attack on the cat) is based on information in A. C. Bent, 1937–38, "Life histories of North American birds of prey," *U.S. National Museum Bulletin* 167:1–409; 170:1–428. A major source on flicker evolution is L. L. Short, 1965, "Hybridization in the flickers (*Colaptes*) of North America," *American Museum of Natural History Bulletin* 129:307–428. A similar discussion of oriole evolution in the Great Plains area is that of C. G. Sibley and L. L. Short, 1964, "Hybridization in the orioles of the Great Plains," *Condor* 66:130–50. A more general discussion of the Great Plains' breeding avifauna is Paul A. Johnsgard, 1979, *Birds of the Great Plains: Breeding Species and Their Distribution*, Lincoln: University of Nebraska Press.

The Prairie. A discussion of the ecology and evolution of eastern and western meadowlarks is in W. E. Lanyon, 1957, "The comparative biology of the meadowlarks (*Sturnella*) in Wisconsin," *Publications of the Nuttall Ornithological Club,* no. 1, pp. 1–67. The breeding biology of dickcissels is discussed by J. L. Zimmerman, 1964, in "Polygyny in the dickcissel," *Auk* 83:534–46; and in J. P. Harmeson, 1974, "Breeding ecology of the dickcissel" *Auk* 91:348–59. An excellent discussion of the biology of deer mice and other mammals is that of C. W. Schwartz and E. R. Schwartz, 1980, *The Wild Mammals of Missouri,* 2d ed. Columbia: University of Missouri Press and Missouri Conservation Commission.

3. *The Future*

The Future of the River. My major source of information of historic changes in the Platte River was G. P. Williams, 1978, "The case of the shrinking channels—the North Platte and Platte Rivers in Nebraska," *U.S. Geological Survey Circular* no. 781. Information on irrigation history and use of the Platte's water was provided by J. M. Jess, 1981, "Surface water use in Nebraska's Platte River Valley," Lincoln: Nebraska Department of Water Resources. Also relevant is the article of B. Vogt, 1978, "Now the river is dying," *National Wildlife* 16(4):4–11; and J. Farrar, 1975, "Crane river," *Nebraskaland,* March, pp. 18–35. Two more technical sources of information on historic changes in Platte flows are R. Bentall, 1982, "Nebraska's Platte River, a graphic analysis of flows," Lincoln: Nebraska Water Survey Paper 53, Conservation and Survey Division, University of Nebraska–Lincoln; and T. R. Eschner, 1981, "Morphologic and hydrologic changes of the Platte River, south-central Nebraska," M.S. thesis, Colorado State University, Fort Collins.

The Future of the People. Information on the history of irrigation and its effects on farming in the Platte Valley is provided by J. L. McKinley, 1939, "The influence of the Platte River upon the history of the valley," Ph.D. dissertation, University of Nebraska–Lincoln (published in mimeo in 1938 by Burgess Publishing Company, Minneapolis, Minn.). More recent information on irrigation acreage, numbers of farms, and irrigation wells in Hall County is from the

"Soil survey of Hall County, Nebraska," U.S. Department of Agriculture Soil Conservation Service and University of Nebraska Conservation and Survey Division, 1957. The most recent (1980) figures were estimated by data provided by the State of Nebraska Department of Agriculture; the Division of Conservation and Survey of the University of Nebraska; and the Nebraska Crop and Livestock Reporting Service, U.S. Department of Agriculture.

The Appendix

Birds. The checklist of birds is based partly on Gary R. Lingle, 1982, "A check-list of the birds of Mormon Island Crane Meadows," *Nebraska Bird Review* 50:27–36; and also on personal data. The final summary of Platte Valley birds is from "The Platte River ecology study" of the U.S. Fish and Wildlife Service, 1981, Jamestown, N. Dak., Special Research Report, Northern Prairie Research Center.

Mammals. The list of mammals of the Platte Valley is mainly derived from range maps by J. K. Jones, Jr., 1964, in "Distribution and taxonomy of mammals of Nebraska," *University of Kansas Publications, Museum of Natural History,* 16:1–356. The Mormon Island records are those of G. Lingle and R. Boner, 1981, "Mormon Island Crane Meadows Management Plan," unpublished report submitted to the Platte River Whooping Crane Habitat Trust, Grand Island.

Reptiles. The list of reptiles from the Mormon Island area is from S. M. Jones, R. E. Ballinger, and J. W. Nietfeldt, 1981, "Herpetofauna of Mormon Island Preserve, Hall County, Nebraska," *Prairie Naturalist* 13:33–41. Other Hall County records are from that paper and from G. E. Hudson, 1958, "The amphibians and reptiles of Nebraska," *Nebraska Conservation and Survey Division Bulletin* 24:1–146.

Fishes. The list of fishes reported from Mormon Island is from J. Cochran and D. Jenson, 1981, "Mormon Island Crane Meadows fish inventory," unpublished report submitted to the Nature Conservancy, Grand Island. Records from Shoemaker Island are from Q. Bliss and S. Schainost, 1973, "Middle Platte Basin stream inventory report," unpublished report, Nebraska Game and Parks Commission,

Bureau of Wildlife Services, Aquatic Wildlife Division. The more general list is partly based on J. Morris, L. Morris, and L. Witt, 1972, "The fishes of Nebraska," Lincoln: Nebraska Game and Parks Commission, with additions and corrections provided by Dr. John Lynch (personal comment).

Plants. The list of plants from the Platte Valley flood plain is based on the list in "The Platte River Ecology Study" of the U.S. Fish and Wildlife Service, with a few name changes or corrections as provided by Dr. Robert Kaul (personal comment).

Acknowledgments

The idea for a short book on the Platte River is one that I have had for several years, although a decision to go ahead and write did not take form until fairly recently. I had originally thought of simply doing some rather unconnected essays on the natural and human history of Shoemaker Island in collaboration with Ms. Phyllis Hobrock, who was intrigued by the homesteading history of that particular locale. However, I soon became caught up in the history of the entire Platte Valley, and thus the focus of the writing changed considerably. While researching the historical aspects I was aided greatly by the library staff of the Nebraska Historical Society, and without the resources of that library I would have had a much more difficult job.

Getting information on the geology and geography of the Platte Valley was made easier by the help of my son, Scott, who provided visual interpretation of aerial photographs, located a variety of maps and other sources of geological information that were vital to my work, and also accompanied me on some of my trips to the area for observations or photographic purposes. Unpublished information on the vertebrates of the Mormon Island Preserve was provided by Gary Lingle of the Platte River Whooping Crane Habitat Maintenance Trust, and additional information on the fishes of the Platte River was offered by Dr. John Lynch of the School of Life Sciences, University of Nebraska–Lincoln. Dr. Robert Kaul, also of the School of Life Sciences, provided advice on various botanical matters. The manuscript was read and criticized by several persons having expertise in various areas, including Scott Johnsgard, Ms. Mary Bomberger, Dr. Mike Voorhies, Mrs. Jean Peterson, and Ms. Carolyn Bantam. I also wish to thank Mr. William Whitney and the University of Nebraska Division of Conservation and Survey for providing photographic assistance.

Index

(This index does not include references to material found in the appendix.)